Progress in Optical Science and Photonics

Volume 34

Indexed by Scopus

The purpose of the series Progress in Optical Science and Photonics is to provide a forum to disseminate the latest research findings in various areas of Optics and its applications. The intended audience are physicists, electrical and electronic engineers, applied mathematicians, biomedical engineers, and advanced graduate students.

Vanita Bhardwaj · Santosh Kumar · Kamal Kishor · Amit Rai
Editors

Optical Fiber Sensors and AI

Exploring the Fusion

Springer

Editors
Vanita Bhardwaj
School of Engineering and Technology
VIPS-TC
New Delhi, Delhi, India

Santosh Kumar
Department of Electronics
and Communications Engineering
Koneru Lakshmaiah Education Foundation
Vaddeswaram, Andhra Pradesh, India

Kamal Kishor
Department of Applied Physics
Delhi Technological University
New Delhi, Delhi, India

Amit Rai
Intelligent Systems Laboratory
Pukyong National University
Busan, Korea (Republic of)

ISSN 2363-5096 ISSN 2363-510X (electronic)
Progress in Optical Science and Photonics
ISBN 978-981-96-6157-2 ISBN 978-981-96-6158-9 (eBook)
https://doi.org/10.1007/978-981-96-6158-9

This Springer imprint is published by the registered company Springer Nature Singapore Pte Ltd.
The registered company address is: 152 Beach Road, #21-01/04 Gateway East, Singapore 189721, Singapore

With deep gratitude, I dedicate my editorial contribution in this volume to my dear friend, **Dr. Sonal Khurana**, *whose unwavering belief in me and relentless encouragement illuminated my path. Without her inspiration, this journey might have remained an unfulfilled dream. Special thanks to her for being a guiding light and a constant source of strength.*

I am also deeply thankful to ***my parents, my children, my husband, and my friends.*** *Their love, support, and faith have been the bedrock of this journey. It is through their constant encouragement that I found the strength and inspiration to bring this book to life. I am also thankful to my fellow editors and authors for their valuable contributions and support throughout the preparation of this volume.*

Preface

The combination of artificial intelligence (AI) and optical fiber sensor technology is changing the way industries and research approach sensing and monitoring. This book, *Optical Fiber Sensors and AI: Exploring the Fusion*, aims to explain how these two powerful technologies work together to create smarter, more efficient sensing systems.

Optical fiber sensors (OFS) are known for their high sensitivity, reliability, and ability to work in many different fields such as healthcare, environmental monitoring, and smart infrastructure. With the addition of AI, these sensors can now process data faster, recognize patterns, and make better predictions, improving their accuracy and performance.

The book is designed to give readers a clear understanding of both the basics and the latest developments in AI-powered optical fiber sensors. It starts by introducing the core principles of optical fiber sensors, then moves into how AI techniques such as machine learning and deep learning are being used to enhance these sensors. It also explores real-world applications and practical case studies, focusing on areas like smart infrastructure, cybersecurity, and hybrid sensor systems. The final chapters look ahead at the future challenges and opportunities in this exciting field.

This book is for researchers, students, and professionals interested in learning about the intersection of optical fiber sensors and AI technologies. It provides easy-to-understand explanations, real-world examples, and valuable insights into how AI is transforming optical fiber sensing. We hope this book will be a helpful resource and inspire further exploration in this rapidly evolving field.

New Delhi, India — Vanita Bhardwaj
Vaddeswaram, India — Santosh Kumar
New Delhi, India — Kamal Kishor
Busan, Korea (Republic of) — Amit Rai

Acknowledgments

We would like to express our sincere thanks to everyone who helped make this book, *Optical Fiber Sensors and AI: Exploring the Fusion*, possible. Without their support, this work would not have been completed.

First, we want to thank the contributors and authors for sharing their expertise and knowledge. Their input has been essential in making this book informative and up-to-date on the topic of combining optical fiber sensors with artificial intelligence.

We also appreciate the guidance and encouragement from our colleagues, mentors, and advisors. Their advice helped shape the content and quality of the book.

A big thank you goes to the team at ***Springer Nature***, who provided excellent support throughout the publishing process. Their work in editing, formatting, and production has been invaluable.

Lastly, we would like to thank our families and friends for their ongoing support and patience while we worked on this book. Their encouragement was a great source of strength.

We hope this book will be helpful to everyone interested in the combination of optical fiber sensors and AI, and that it will inspire more research and developments in this exciting area.

About This Book

This book highlights the exciting developments in optical fiber sensors and how artificial intelligence (AI) is boosting their performance and applications. It starts with an easy-to-understand introduction to the basics of optical fiber sensors and their many uses. Then, it moves on to the latest technological advancements, showing how AI is making these sensors smarter and more efficient. The book contains chapters demonstrating how machine learning contributes to real-time data analysis and how deep learning enhances sensor systems. There is also a focus on designing better sensor networks with the help of AI. The book explains how combining AI with the Internet of Things (IoT) and optical fiber sensors can create smart infrastructure solutions. Real-world case studies illustrate how AI-enhanced fiber optic sensors are benefiting fields like healthcare and environmental monitoring. The book wraps up with a look at future trends and challenges in the world of AI-powered optical fiber sensing. This book is perfect for researchers, engineers, and anyone interested in the powerful combination of AI and optical fiber technology. It provides valuable insights into how these technologies can work together to create innovative and practical solutions.

Contents

Introduction of Optical Fiber: Fundamentals and Applications

Rekha Bhardwaj, Garima Chanana, Sonal Khurana, and Navneet Kumar

Abstract The unique features of fiber optics have been helpful in its massive application across several domains for fast and long-distance data transfer in modern communication. This chapter looks into the basics of the operation of fiber optics along with its fabrication and design techniques. We further discuss the diverse applications of fiber optics, ranging from medical imaging and industrial sensing to secure military communications and renewable energy solutions. Furthermore, the future trends of optical fiber technologies are examined. With continuous advancements, optical fiber technology is assisting in shaping the future of global communication and automation. The increasing usage of sensors in everyday tasks generates important information for medical diagnostics in the healthcare field that in turn influences the standard of life. Moreover, these sensors are famous for their extraordinary linear response, quick response times for real-time analysis, and essential variation to changes in the outside environment. The vital characteristics of optical fiber sensors compel them particularly important for various uses.

Keywords Fiber optics · Optical communication · High-speed data transmission · Optical fiber applications

1 Introduction

Fiber optics is a groundbreaking technology that has revolutionized the way information is transmitted and accessed in the modern world [1]. The basic working principle of fiber optics is transmission of light through flexible, transparent fibers which are usually made using silica or plastic material. These fibers help in achieving high-speed and long-distance communication using light [2]. One of the peculiar properties

R. Bhardwaj
Department of Applied Sciences, Bharati Vidyapeeth's College of Engineering, New Delhi, India

G. Chanana (✉) · S. Khurana · N. Kumar
School of Engineering and Technology, Vivekananda Institute of Professional Studies - Technical Campus, New Delhi, India
e-mail: garima.chanana@vips.edu

V. Bhardwaj et al. (eds.), *Optical Fiber Sensors and AI*, Progress in Optical Science and Photonics 34, https://doi.org/10.1007/978-981-96-6158-9_1

of optical fibers is their ability to carry large amounts of data with minimal signal loss along with immunity to external electromagnetic interference. These properties have positioned them as a cornerstone for telecommunications, healthcare, industrial automation, and military applications [3]. The origins of fiber optics date back to the nineteenth century, when physicist John Tyndall [4] demonstrated the principle of total internal reflection, a phenomenon that forms the foundation of optical fiber technology. Yet it was in the twentieth century that the possibility of sending light along thin fibers became truly appreciated. The 1960s saw the development of LASERS [5] as powerful light sources. This resulted in the production of low-loss optical fibers which led to numerous developments in fiber optics [6].

Optical fiber sensors provide many advantages in comparison to other sensing technologies for various purposes with essential possibilities in measuring tasks. Optical fibers provide various advantages for sensing purposes, especially their compact form factor, the requirement to eliminate electric energy at far-off location, and their capability to divide various sensors which includes the fiber's length. This distribution could be achieved by the usage of light wavelength variations for individual sensors or by measurement of the time delay of light as it does communication with each sensor present in the optical fiber. Therefore, the purposive applications vary from risky circumstances which comprise chemical, and varied industry issues to more distinctive and simple implementations. Still, the development in determined optical fiber sensors, example outlined for detecting corrosion, is yet in its developing stage. Optical fiber sensors can be categorized on the basis of varied factors. These sensors are often assembled based on sensor position inside the fiber, in addition to utilization. In the situation of use cases, optical sensors can be grouped on the basis of the category of functions they are proposed to estimate, in particular, temperature sensors, pressure sensors, chemical sensors, biosensors, intensity, and polarization-based sensors. Accordingly, utilizing the sensor, the optical fiber sensors can be grouped as intrinsic or extrinsic.

Intrinsic sensors in optical fiber are the sensors which utilize the optical fiber as the sensing element far apart from extrinsic fiber which act as a channel for carrying light signals to and from an external sensing element, intrinsic fiber optic sensors depend on modifications in the attributes of the optical fiber only for detecting and measuring to detect and measure varied physical parameters. Based on the sensing location (i.e., changes in sensing of light rays inside or outside the optical fibers), sensors are categorized into two types: intrinsic and extrinsic optical fiber sensors. In extrinsic optical fiber sensors, the light ray changes due to the optical device whereas in intrinsic optical fiber light ray changes in optical fiber itself only. Extrinsic optical fiber sensors are commonly used as temperature sensors, pressure sensors, etc. and intrinsic optical fiber sensors are used as vibration sensors, rotation sensors, etc.

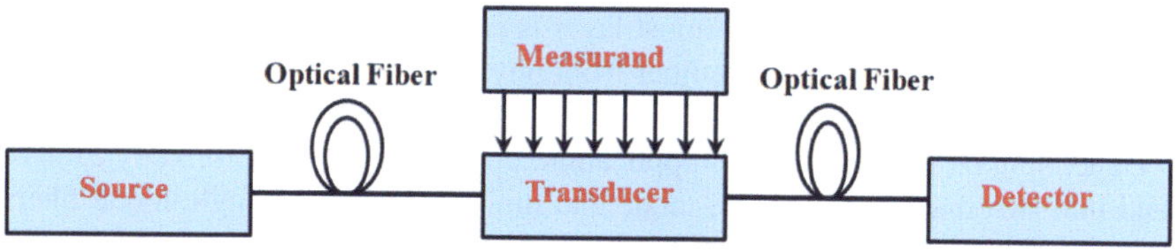

Fig. 1 Schematic representation of optical fiber sensors

2 Fundamentals of Optical Fiber Sensors (OFS)

OFSs are able to monitor almost all physical parameters and lots of chemical parameters too. For the chemical parameter detection, OFSs need some functional material coating to enhance the device performance [7]. In OFS chemical sensors and electromagnetic radiation is used to generate analytical signals in a transduction element. The optical fiber sensor consists of sensing element coupled with transduction element, and the receptor is used to identify various chemical parameters. The main components of the optical fiber sensor are optical source, optical fiber, sensing element, and optical detector described in Fig. 1. OFSs can be categorized into three main groups on the basis of the optical parameters' (phase, amplitude, and state of polarization) modulation.

More light is needed in intensity-modulated optical fiber sensors as compared to phase-modulated sensors. Thus, optical fibers with large core diameter are generally used in intensity-modulated optical fiber sensors. Phase-modulated sensors are also known as interferometric sensors. They require an optical splitter and coupler for splitting and coupling of light beams coming from the coherent source. This splitting and coupling of light cause shifts in phase detected by interferometer. The result obtained from the phase-modulated sensor is more accurate than the intensity-modulated OFSs [5].

3 Working Principle of Optical Fiber

The primary operation of a working fiber optic cable is based on total internal reflection. In this phenomenon, when light travels within a medium having a higher refractive index and meets a boundary with a lower refractive index, it reflects totally within the medium. For this reason, optical fibers have been designed to use this principle to transmit light efficiently along the length of the cable [7]. An optical fiber has three main parts, which are the core, cladding, and coating. Core is the central part through which light is transmitted. The refractive index and diameter of the core determine the mode of operation of the fiber and its ability to carry light signals. There is another layer known as cladding, which surrounds the core, but it has a lower refractive index than that of the core. This ensures that the light remains confined to the core through total internal reflection. The outermost layer is called the coating which is used as a protective layer that shields the fiber from physical damage, moisture, and other environmental degrading factors.

Depending on their design and application, optical fibers can be broadly categorized into two main types: single-mode and multimode fibers. Single-mode fibers have a small core diameter and support the propagation of just one mode of light [8]. These are useful for long-distance and high-bandwidth communication. Multimode fibers [9] have a larger core and support multiple light modes within the fiber. They are usually employed in short-distance applications [10, 11].

The fiber optical cables have found diverse applications across numerous fields, which demonstrates its versatility and technological significance [12]. Fiber optic cables form the backbone of modern communication networks, facilitating Internet and television services. In healthcare [13], fiber optic sensors are used in endoscopy, laser surgeries, and advanced imaging techniques which enable minimally invasive surgical procedures. Optical fibers are also employed for sensing temperature [14], pressure [15], and strain [16] for improving efficiency and safety in industrial automation. They are also used for secure and high-speed communication and navigation systems for military [17] and for aerospace applications [18, 19]. They also enable precision measurements and high-resolution imaging in fields like astronomy [20, 21].

4 Fabrication and Design

The fabrication and design of optical fibers is a highly intricate process which determines the performance and functionality of fiber optic cable [22, 23]. The fibers are fabricated using either glass or polymer materials, with silica glass being the most commonly used due to its superior optical properties such as low attenuation and high mechanical strength [24, 25]. The fabrication of optical fibers begins with the creation of a preform [26, 27]. It is a solid cylindrical structure made of highly pure glass. The preform acts as an initial blueprint for the optical fiber. Using the preform, various chemical vapor deposition (CVD) methods such as modified chemical vapor deposition [28], outside vapor deposition [29], and vapor phase axial deposition [30] are widely employed for depositing layers of silica and dopants to form the core and cladding of the fiber. This ensures precise control of refractive indices. Once the preform is prepared, it is heated in a high-temperature furnace, and the softened glass is drawn into thin fibers. The drawing process needs precise control over temperature, tension in the cable, and speed to maintain uniformity in diameter. The diameter is typically around 125 μm for standard fibers. A protective polymer coating is applied immediately after drawing to enhance the fiber's mechanical strength and environmental resistance [31].

The choice of materials significantly influences the fiber cable's optical and mechanical properties [32]. Silica is doped with materials like germanium, phosphorus, or boron to adjust the refractive index of the core. One of the popular optical fibers used is polymer fibers [33] which offers flexibility and lower cost but are limited in terms of performance compared to silica-based fibers [34]. Advances in material science have led to the development of specialty fibers, such as fluoride [35] or chalcogenide [36] glass fibers, which operate in mid-infrared ranges, expanding their applicability.

Micro-structured optical fibers have emerged as an innovative design in fiber optics. These fibers incorporate air holes in their cladding, allowing for unique light-guiding properties. They are particularly useful for sensing applications [37]. Photonic crystal fibers are another type of optical fibers which is useful for high nonlinear and single-mode operation with extended wavelength range applications [38].

In the various types of optical fibers discussed above, the key idea is that the core and cladding of an optical fiber play pivotal roles in light transmission. Fabrication and design of these parts of optical fiber is important for specialized applications. However, there are some fabrication challenges also such as achieving high purity levels, minimizing defects, and reducing attenuation. Innovations like nanostructured coatings [39] and advanced doping techniques are used to address these challenges. It is also important to ensure the quality of optical fibers by rigorous testing of parameters such as tensile strength, attenuation, dispersion, and environmental stability. For this purpose, advanced testing equipment and methodologies are used to identify any manufacturing defects and ensure compliance with industry standards.

5 Classification of Optical Fiber Sensors

There are various classifications of optical fiber sensors that could be recognized as follows: In the middle of the earliest optical fiber sensors developed were intensity-modulated sensors. These sensors are adept at detecting physical alterations or disturbances in incoming light, including attenuation and transverse electromagnetic losses. Their advantages lie in their simplicity and lower cost applications. This necessitates the implementation of a reference system to address the problem [31]. The Fiber Bragg Grating (FBG) sensor is particularly notable, as it is widely used across various applications for its ability to conduct both single-point and multi-point sensing. The principle of interferometry is employed by phase-modulated sensors to analyze the interference of light transmitted through fiber optics. These sensors are preferred for their exceptional sensitivity and precision, which, however, contributes to their elevated cost. Prominent examples of phase-modulated sensors include the Michelson, Sagnac, Fabry–Perot, and Mach–Zehnder interferometers. Optical time domain reflectometry is a method where pulse of light is sent through the optical fiber, and reflections from imperfections such as temperature and strain changes are measured [32]. By analyzing the time it takes for the light to return, OTDR can detect faults, measure distances, and monitor physical parameters along with the fiber. Sensors that are based on polarization principles can discern variations in light that arise from changes in the state of polarization. They utilize the birefringence effect within optical fibers, where the refractive index is contingent upon the polarization state [40, 41].

6 Applications of Fiber Optics

Fiber optics are being used in diverse fields which range from telecommunications [40] to medical applications, industrial automation [42], and military systems [43]. The primary and most widespread application of fiber optics is in telecommunications. Fiber optic cables are the backbone of modern high-speed Internet, enabling faster data transmission over long distances with minimal signal loss. They are essential in transmitting voice, video, and data signals for broadband networks, undersea cables, and satellite communication. Due to their high bandwidth and reliability, fiber optics support the increasing demand for 5G networks and other emerging technologies.

6.1 Medical Field

Fiber optics play a significant role in healthcare, particularly in diagnostics and minimally invasive surgeries. Endoscopes [44], which use fiber optics to visualize internal organs, enable doctors to diagnose and treat conditions with precision. Fiber optics are also used in laser surgeries, phototherapy, and biomedical sensors for monitoring physiological parameters like oxygen levels and blood flow. Optical coherence tomography (OCT) [45], a technique that provides high-resolution imaging of tissues, relies on fiber optics to diagnose eye diseases, cardiovascular conditions, and cancer.

6.2 Sensor

In industries, fiber optics are used for sensing, monitoring, and automation. Optical fibers serve as sensors in harsh environments to measure temperature, pressure, and strain, especially in oil and gas exploration and manufacturing [46]. Their ability to withstand extreme conditions makes them ideal for applications in power plants, aerospace, and civil engineering projects. Fiber optics are also used in robotics [47] for precise control and in 3D printing to improve laser systems.

6.3 Military

Fiber optics are critical in military applications due to their secure communication capabilities. They are used in avionics, navigation systems, and missile guidance systems, providing resilience to electromagnetic interference and enhanced security. Military vehicles and submarines use fiber optic networks to transmit data for surveillance, radar systems, and real-time communication.

6.4 Energy and Communication Sector

Fiber optics are used in the energy sector for monitoring power lines, wind turbines, and pipelines [48]. They help in detecting faults, measuring strain, and ensuring operational efficiency. In solar power systems, fiber optics guide sunlight to generate energy efficiently. In summary, fiber optics have revolutionized countless industries, offering unmatched speed, precision, and reliability. From telecommunications to medical advancements, aerospace, and even environmental monitoring, the versatility of fiber optics continues to drive innovation and efficiency. As technology evolves, their applications will only expand further, shaping the future of communication, automation, and sustainability across the globe.

7 Advantages of Optical Fiber

The adoption of fiber optics has been driven by its numerous advantages over traditional transmission media, such as copper wires. Optical fibers can carry significantly more data compared to conventional cables, making them indispensable in high-speed networks [10]. With advancements in fiber materials and new techniques of manufacturing, modern optical fibers show minimal attenuation which helps in efficient long-distance communication [11]. Unlike copper wires, optical fibers are not affected by electromagnetic fields as signal transmission occurs using light instead of current. This helps in ensuring reliable data transmission. Optical fibers are much lighter and thinner than copper cables, simplifying installation and reducing infrastructure costs. They are resistant to corrosion and environmental factors and hence are well suited for harsh environments.

7.1 Long-Range Transmission

Long-range transmission refers to ability of transfer of signals without any distortion, quality of the signal, and strength of the data across the thousand kilometer. During the transmission of data there are some key parameters during transmission of signal attenuation, noise and dispersion must be less because these factors degrade the quality of the signals [9].

7.2 Low Loss Fiber

There are two modes of transmission used in optical fiber: single-mode optical fiber and multimode fiber. Single-mode fiber is generally used for long-range communication because of lower dispersion and attention as compared to multimode optical fiber.

7.3 Optical Amplifier

There are many optical amplifiers (Raman optical amplifier) which are used to boost signals for long-distance transmission. Optical amplifiers are placed at the intervals of optical fiber without any conversion of electrical form, which helps to transmit data for long distances.

7.4 Safe in Hazardous Environments

For transmission of data, optical fiber uses light instead of radiation and electricity. Due to the use of light, it is suitable for flammable gases, explosive gas areas, and no risk of electric shocks and sparking.

7.5 High Bandwidth

Optical fiber has high bandwidth which means transfer of large amounts of data through a single cable at high speed. Optical fiber transfers data faster as compared to electrical signals.

7.6 Multiplexing Division of Wavelength

Transfer of multiple data through single optical fiber using Wavelength Division Multiplexing (WDM). WDM is a way to transmit data through different wavelengths, each wavelength acting as a separate channel for transmission of data.

7.7 Low Signal Attenuation

Low signal attenuation means minimum loss of signal strength and attenuation while light transmits through cable. In optical fiber, attenuation is caused by the scattering, absorption, and bending losses.

7.8 Compact and Lightweight

Optical fibers are generally made of plastic or silica which are much lighter than the other metal or traditional cables. Optical fiber takes less space as compared to copper cable during underground installation, server room, etc.

7.9 Long Transmission Scalability

Due to low attenuation in optical fiber, we can transfer data over a thousand kilometers. For transmission of data, repeaters and amplifiers are added without changing the cable. Optical fiber can be used in submarines for international data traffics.

7.10 Emerging Technologies or Modular Upgrades

Addition of channel or wavelength can be introduced in the optical fiber which increases the capacity and strength of the optical fiber.

7.11 Cost-Effective

In the initial stage, installation of optical fiber becomes more. Due to long durability, minimum maintenance, and high energy efficiency, this technology becomes cost-effective in long term.

8 Limitations of Optical Fiber

8.1 *Signal Loss in Optical Fiber*

Optical fibers are made up of plastic and silica which make optical fiber fragile and bending and stretching of optical fiber cause signal loss due to attenuation.

8.2 *Limitation in Power Transmission*

Optical fiber transmits data through light, but optical fiber cannot transmit electric power like copper wire. Due to this reason transmission of power and data requires separate cable, for example, power over Ethernet.

8.3 *Compatibility Issues*

For transmission of data through optical fiber, the installation of specialized equipment or hardware like converter, receiver, etc. is required.

8.4 *Amplifier Dependency*

In multimode, optical fibers suffer from modular dispersion due to this transmission of data for long-distance optical fiber periodic amplifiers.

9 Future Trends in Optical Fiber

The future of fiber optics is poised for exponential growth, driven by the ever-increasing demand for faster, more reliable communication. Emerging technologies, such as 5G networks [20] and the Internet of Things (IoT), are expected to rely heavily on fiber optics for seamless connectivity and transmission. In the context of global sustainability, fiber optics offers an environmentally friendly and cost-effective alternative for efficient management of infrastructures [32]. With continued advancements in fiber optic materials, manufacturing techniques, and network design, fiber optics is set to remain a pivotal technology in shaping the future of communication and innovation.

9.1 Integration with Quantum Communication

Using quantum encryption technique optical fiber plays an important role in quantum communication systems for developing the central quantum network and other networks.

9.2 5G and Beyond

Optical fiber has low latency, dense network, and minimal signal degradation which make real applications like IOT, autonomous vehicles, etc. Dense network of an optical fiber is required for future 5G or 6G networks.

9.3 Automation and AI in Network Management

By use of AI and machine learning in optical fiber networks, optical fiber detects their own fault and damage and heals the optical network.

9.4 Advancement in Undersea Cable

Use of AI and machine learning technology develops the new design and more advanced optical fiber by adding the more advanced amplifier to support growing global data traffic.

9.5 Precession and Enhanced Sensitivity in Optical Fiber

The advancement of nanotechnology enables a highly enhanced optical fiber sensor for precise detection of pressure, temperature, and chemical composition.

9.6 Integration with IoT and Smart Systems

Optical fiber sensor will integrate IoT systems which are helpful for development of smart cities, homes, and industries. Optical fiber sensors will also help to enable the critical systems like oil and power grids.

9.7 Healthcare Application:

In future, optical fiber cable can be used in biomarkers for disease detection, in endoscopic tools, and for monitoring and measuring blood oxygen levels, etc. in real time.

10 Conclusion

Optical fiber sensors provide many advantages over various types of sensors. These benefits are closely tied to the distinctive features of optical fibers, including their compact form factor, lightweight design, durability in extreme temperatures and pressures, and their immunity to electromagnetic interference, between other factors. The analysis of light properties facilitates sensing, provided by the calculations of variables, for example, temperature, pressure, and strain. Optical fiber technology was mainly created for applications in data networking and electronic communications. Despite this, sensors depending on optical fiber have progressed very fast because of their special sensing abilities and capacity to work in diverse and difficult conditions. An in-depth summary of the sensing principle, production functionality, utilization, pros, and cons of different fiber optic sensors is presented. Current developments in different sensing areas, having environmental, industrial, and biomedical, are analyzed for every classification of optical fiber sensor. This highlights the process and techniques employed to address the sensing challenges.

References

1. W.B. Allan, *Fiberoptics: Theory and Practice*. Springer Science & Business Media (2012)
2. M. Elsherif, A.E. Salih, M.G. Muñoz, F. Alam, B. AlQattan, D.S. Antonysamy, H. Butt et al., Optical fiber sensors: working principle, applications, and limitations. Adv. Photon. Res. **3**(11), 2100371 (2022)
3. J. Hecht, Understanding fiber optics. Jeff Hecht (2015)
4. J. Tyndall, Six lectures on light: delivered in the United States in 1872–1873. Longmans, Green (1895)
5. T.H. Maiman, Stimulated optical radiation in ruby (1960)
6. G.H. Sigel Jr., D.C. Tran, Ultra-low loss optical fibers: an overview. Infrared Optical Mater. Fibers III **484**, 2–6 (1984)
7. S.G. Foord, J. Lees, Special issue paper. Principles of fibre-optical cable design, in *Proceedings of the Institution of Electrical Engineers* (Vol. 123, No. 6, pp. 597–602). IET Digital Library (1976)
8. L.B. Jeunhomme, *Single-Mode Fiber Optics: Prinicples and Applications*. Routledge (2019)
9. N. Borhani, E. Kakkava, C. Moser, D. Psaltis, Learning to see through multimode fibers. Optica **5**(8), 960–966 (2018)
10. V.W. Chan, K.L. Hall, E. Modiano, K.A. Rauschenbach, Architectures and technologies for high-speed optical data networks. J. Lightwave Technol. **16**(12), 2146 (1998)

11. F. Yaman, N. Bai, B. Zhu, T. Wang, G. Li, Long distance transmission in few-mode fibers. Opt. Express **18**(12), 13250–13257 (2010)
12. P.J. Winzer, D.T. Neilson, A.R. Chraplyvy, Fiber-optic transmission and networking: the previous 20 and the next 20 years. Opt. Express **26**(18), 24190–24239 (2018)
13. A.G. Leal-Junior, C.A. Diaz, L.M. Avellar, M.J. Pontes, C. Marques, A. Frizera, Polymer optical fiber sensors in healthcare applications: a comprehensive review. Sensors **19**(14), 3156 (2019)
14. S. Ma, Y. Xu, Y. Pang, X. Zhao, Y. Li, Z. Qin, ...X. Bao, Optical fiber sensors for high-temperature monitoring: a review. Sensors **22**(15), 5722 (2022)
15. É. Pinet, Pressure measurement with fiber-optic sensors: commercial technologies and applications, in 21st *International Conference on Optical Fiber Sensors* (Vol. 7753, pp. 51–54). SPIE (2011)
16. C.M. Lawrence, D.V. Nelson, E. Udd, T. Bennett, A fiber optic sensor for transverse strain measurement. Exp. Mech. **39**, 202–209 (1999)
17. P.B. Ruffin, Review of fiber optics technology for military applications. Novel Mater. Crystal Growth Techniques Nonlinear Opt. Dev. Critical Rev., **10299**, 2–25 (2000)
18. A. Rovera, A. Tancau, N. Boetti, M.D. Dalla Vedova, P. Maggiore, D. Janner, Fiber optic sensors for harsh and high radiation environments in aerospace applications. Sensors **23**(5), 2512 (2023)
19. W.D. Heacox, P. Connes, Optical fibers in astronomical instruments. Astron. Astrophys. Rev. **3**, 169–199 (1992)
20. S. Miladić-Tešić, G. Marković, D. Peraković, I. Cvitić, A review of optical networking technologies supporting 5G communication infrastructure. Wireless Netw. **28**(1), 459–467 (2022)
21. M.D.F.F. Domingues, A. Radwan, M.D.F.F. Domingues, A. Radwan, Optical fiber sensors in IoT (pp. 73–86). Springer International Publishing (2017)
22. P. Kot, K.S. Hashim, M. Muradov, R. Al-Khaddar, How can sensors be used for sustainability improvement? In Methods in Sustainability Science (pp. 321–344). Elsevier (2021)
23. X. Wang, Q. Nie, T. Xu, L. Liu, A review of the fabrication of optic fiber. ICO20: Optical Design and Fabrication, **6034**, 346–354 (2006)
24. D.J. DiGiovanni, S.K. Das, L.L. Blyler, W. White, R.K. Boncek, S.E. Golowich, Design of optical fibers for communications systems. Optical Fiber Telecommun. **IV-A**, 17–79 Golowich
25. K. Oh, U.C. Paek, *Silica Optical Fiber Technology for Devices and Components: Design, Fabrication, and International Standards* (John Wiley & Sons, 2012)
26. Y. Chu, X. Fu, Y. Luo, J. Canning, Y. Tian, K. Cook, G.D. Peng et al., Silica optical fiber drawn from 3D printed preforms. Opt. Lett. **44**(21), 5358–5361 (2019)
27. T. Oriekhov, C.M. Harvey, K. Mühlberger, M. Fokine, Specialty optical fiber fabrication: preform manufacturing based on asymmetrical CO laser heating. JOSA B **38**(12), F130–F137 (2021)
28. K.S. Kim, Modeling and analysis of modified chemical vapor deposition of optical fiber preforms. Chem. Eng. Sci. **44**(11), 2475–2482 (1989)
29. M. Blankenship, C. Deneka, The outside vapor deposition method of fabricating optical waveguide fibers. IEEE J. Quantum Electron. **18**(10), 1418–1423 (1982)
30. H. Murata, Recent developments in vapor phase axial deposition. J. Lightwave Technol. **4**(8), 1026–1033 (1986)
31. Y. Jaluria, Y. Jaluria, Manufacture of optical fibers: drawing and coating processes. Advanced Materials Processing and Manufacturing, pp. 239–286 (2018)
32. J. Ballato, P. Dragic, Materials development for next generation optical fiber. Materials **7**(6), 4411–4430 (2014)
33. R. Kotek, Recent advances in polymer fibers. Polym. Rev. **48**(2), 221–229 (2008)
34. Y. Yamamoto, Y. Tamura, T. Hasegawa, Silica-based highly nonlinear fibers and their applications. SEI Tech. Rev **83**, 15–20 (2016)
35. I.D. Aggarwal, G. Lu (Eds.). (2013). *Fluoride Glass Fiber Optics*. Academic Press

36. J. Nishii, S. Morimoto, I. Inagawa, R. Iizuka, T. Yamashita, T. Yamagishi, Recent advances and trends in chalcogenide glass fiber technology: a review. J. Non-Cryst. Solids **140**, 199–208 (1992)
37. Z. Liu, Z.F. Zhang, H.Y. Tam, X. Tao, Multifunctional smart optical fibers: materials, fabrication, and sensing applications, in *Photonics* (Vol. 6, No. 2, p. 48). MDPI (2019)
38. P. Russell, Photonic crystal fibers. science **299**(5605), 358–362 (2003)
39. C.R. Zamarreno, I.R. Matias, F.J. Arregui, Nanofabrication techniques applied to the development of novel optical fiber sensors based on nanostructured coatings. IEEE Sens. J. **12**(8), 2699–2710 (2012)
40. A. Willner, *Optical Fiber Telecommunications* (Vol. 11). Academic Press (2019)
41. A. Mendez, Medical applications of fiber-optics: optical fiber sees growth as medical sensors. Laser Focus World **1**(01) (2011)
42. T. Bangemann, S. Karnouskos, R. Camp, O. Carlsson, M. Riedl, S. McLeod, ...P. Stluka, State of the art in industrial automation. Industrial Cloud-Based Cyber-Physical Systems: The IMC-AESOP Approach, pp. 23–47 (2014)
43. G. Wild, Novel photonics sensing and systems for military metrology, in *2015 IEEE Metrology for Aerospace (MetroAeroSpace),* pp. 129–134. IEEE (2015)
44. P. Murphy, A fibre—optic endoscope used for nasal intubation. Anaesthesia **22**(3), 489–491 (1967)
45. E. Regar, N. Gonzalo, G. Van Soest, P.W. Serruys, Optical coherence tomography. Cardiovascular Catheterization and Intervention, 388–403 (2010)
46. B.E. Jones, Optical Fibersensors and systems for industry. J. Phys. E: Sci. Instrum. **18**(9), 770 (1985)
47. B. Culshaw, Fiberoptic sensors in advanced robotic systems, in *Sensors and Sensory Systems for Advanced Robots*, pp. 243–257. Berlin, Heidelberg: Springer Berlin Heidelberg (1988)
48. T. Bosselmann, M. Willsch, W. Ecke, The rising demand for energy: a potential for optical fiber sensors in the monitoring sector, in *Smart Sensor Phenomena, Technology, Networks, and Systems 2008* (Vol. 6933, pp. 122–129). SPIE (2008)

Advancements in Optical Fibre Sensors Using Artificial Intelligence Technology

Rekha Bhardwaj, Medha Bhushan, Sonal Khurana, and Ranjana Jha

Abstract Over the last three decades, demand of smart and intelligent society increases day by day and use of artificial technique with optical fibre sensor plays pivotal role for the development of smart cities. Optical fibre sensors have large number of applications in the field of health monitoring, military, cybersecurity, energy, aerospace, etc. However, sensors are made up of optical fibre, for excellent accuracy and sensing performance optical fibre sensor merges with artificial intelligence technology for better accuracy and performance. Optical fibre sensor technology is the backbone for future smart and intelligence city. But there is some limitations and boundaries in optical fibre sensor technology like reduction of signal-to-noise ratio, cross sensitivity, large data generation, low speed, etc. These boundaries are overcome by merging optical sensor with artificial intelligence and machine learning. In this chapter, we discuss about compressive study of optical fibre and its types and their applications and limitations. This chapter also highlights how artificial intelligence and its subset machine learning technology enhance the accuracy, performance of sensor and new applications for the development of smart and intelligence society.

Keywords Artificial intelligence · Sensors · Machine learning · Optical fibre

R. Bhardwaj (✉)
Department of Applied Sciences, Bharati Vidyapeeth's College of Engineering, New Delhi, India
e-mail: rekhabhardwaj488@gmail.com

M. Bhushan
Department of Basic Sciences, G. Narayanamma Institute of Technology and Science, Shaikpet, Hyderabad, India

S. Khurana
Department of Engineering, Vivekananda Institute of Professional Studies, New Delhi, India

R. Jha
Research Lab for Energy System, Department of Physics, Netaji Subhas University of Technology, New Delhi, India

V. Bhardwaj et al. (eds.), *Optical Fiber Sensors and AI*, Progress in Optical Science and Photonics 34, https://doi.org/10.1007/978-981-96-6158-9_2

1 Introduction

Our world is moving towards powerful computing tools like cloud computing, digital revolution, etc., which consist of smart devices like hardware, sensor (Alexa, Google Home, etc.), Internet of Things (IoT), Artificial Intelligence, etc. Using these technologies, we can easily access large data in few minutes. This digital revolution has generated new era like high-performance computer, analytic technology, development of large number of data and their storage, sensing hardware and the latest breakthrough in artificial intelligence and machine learning in nowadays [1]. This encourages the research and development for major breakthrough for evolution of efficient machine learning algorithms. These machine learning algorithms, continuously forwarded to development of large-scale data analysis, help in solving specific problem and aid in neural networks. Interpretation of conventional machine learning algorithms is based on the individual selection of specialized features; meanwhile, deep neural network is one of the automatic generated features from the machine learning process [2].

Artificial intelligence technology performance increases day by day as compared to practical task done by human-like machine translation, speech recognition, etc. [3]. Artificial intelligence sensors are the fourth industrial revolution which is founded by industrial Internet of Things. Smart sensors are the major components of any modern instruments or device. In our twenty-first century, all the devices have at least one sensor or many sensors are immutable in device. These types of sensors are found in practical applications like civil engineering, medical diagnostics, environment monitoring, agriculture framing, fault detection, etc. Sensors are more accurate and sophisticated as Raman spectrometers because it measures individual molecules, vibrations, concentration of solution and chemicals, physical parameters, etc. The Introduction of fundamental function of sensor is to sense or detect more accurately and depending on the types of sensors its outcomes may be distinct or different from digital image, gas sensor and electrical voltage signals [2]. Most of the sensors are made up of optical fibre sensors because optical fibre sensors are more accurate, light weight and have ability to immune in electromagnetic interference. Optical fibre technologies offer to measure parameters in harsh environment, vibrational parameters, temperature, mechanical, low energy consumption, etc., where other electrical sensors are not able to measure the physical parameters. Due to these unique properties optical fibre technology is applied to machine learning algorithms for unique advancement for large data storage, wavelength division technology, give specific inputs for algorithm, etc.

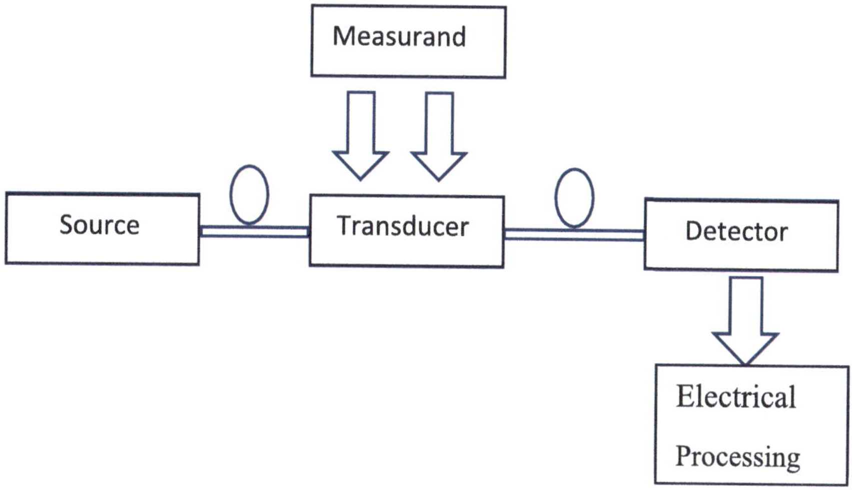

Fig. 1 Block diagram of optical fibre sensor

2 Basics of Optical Fibre Sensor

A sensor which used optical fibre for detecting element is known as optical fibre sensor. Optical fibre sensors are compact in size and its doesn't require any electrical supply for detecting element. Optical fibre sensor depends upon the length of the fibre and wavelength of light. Each sensor has different wavelength and length of the fibre that depend upon the type of the sensor. Optical fibre sensors are electromagnetically resistant so they do not conduct electricity [4].

2.1 Components of Optical Fibre

Major components of optical fibre sensors are LED, laser diode, optical detector, electronic processing and sensing element like transducer. Block diagram of optical fibre sensor is displayed in Fig. 1.

2.2 Working Principle of Optical Fibre Sensor

Working of optical fibre sensor depends upon the transducer, because transducer changes the parameters of optical fibre like intensity, phase, wavelength, polarization, etc. Optical fibre sensor works on the principle of super-luminescent source called laser or light, which can be transmitted through the fibre and then these light sources reach to detector which measure the changes [5].

2.3 Key Components of Optical Fibre

(a) **Sensing Mechanism**: Optical fibre sensor mechanism is based on the interaction of light between the core and environment. All the sensors depend on the basis of characteristics of light wavelength, polarization, intensity, etc. These variations help to measure temperature, strain, pressure, etc.
(b) **Phase Sensing**: When light travels through optical fibre, phase of light changes with length of fibre because of external conditions like temperature, strain, pressure, etc. Using interferometric technique change in phase of light is measured.
(c) **Wavelength Sensing**: Fibre Bragg grating works on the principle of periodic structure which is inscribed in the optical fibre structure. Due to external weather condition spacing between the grating changes. This type of sensor is generally used in health monitoring.
(d) **Polarization Sensor**: During the travelling of light optical fibre state of a polarization changes through external behaviour of environment. Electromagnetic fields need to measured.

3 Types of Optical Fibre Sensor on the Basis of Transducing Approach

On the basis of sensing location, optical fibre sensor is divided into three types:

(i) Intrinsic fibre optical sensor.
(ii) Extrinsic fibre optical sensor.
(iii) Hybrid fibre optical sensor.

3.1 Intrinsic Fibre Optical Sensor

In intrinsic fibre optical sensor, detection of element occurs within the fibre itself. Intrinsic fibre optical sensor depends upon the characteristics features of optical fibre sensor to change the environment condition to modulate the signal of light through it. The main features of sensors are to detect material from long-range distance because it depends upon the light physical properties like frequency, phase, intensity, polarization, etc. [2] (Fig. 2).

3.2 Extrinsic Fibre Optical Sensor

Extrinsic fibre optical sensor transmits information through black box, this black box is designed using mirror, gas, or using other devices which produces optical signal.

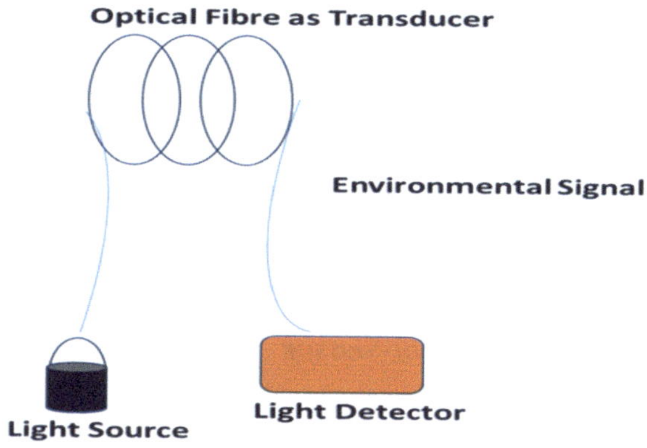

Fig. 2 Intrinsic fibre optical sensor

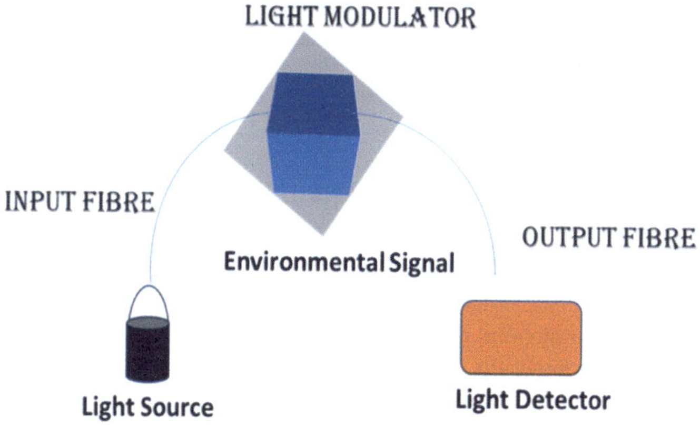

Fig. 3 Extrinsic fibre optics sensor

This sensor is generally used to measure velocity, revolution, torque, acceleration, etc. This sensor is also used to measure the temperature of aircraft engine. Extrinsic fibre optical sensor needs protection from noise corruption during measurement of signal [6] (Fig. 3).

3.3 *Hybrid Fibre Optical Sensor*

In this type of sensor, optical fibre works as a front-end component in device. Hybrid fibre optical sensor has widespread use in analyses of simultaneous transmission and sensing capabilities. Hybrid fibre optical sensor reduces transmission loss and also has long-term reliability and less maintenance cost as compared to electronic device [7].

4 On the Basis of Topology Optical Fibre Sensor Divided into Three Types

(i) **Single-Point sensor.**
(ii) **Multi-point (quasi-distributed) sensor.**
(iii) **Distributed sensor.**

4.1 Single-Point Sensor

Single-point optical fibre is employed to sense only specific parameters like temperatures, pressures, strain, etc. [8]. Single point has high durability, small size, high sensitivity, high sensitivity and more accuracy because it senses at a fixed point or single parameter at a time. Single-point sensor is used in medical application, light transmissions and parameter's interaction [9].

4.2 Multi-Point (Quasi-Distributed) Sensor

Multi-point sensor has several points to measure discrete parameters in a single device. Multi-point sensor is also called quasi-distributed sensor [10]. In this type of sensor, number of sensing point increases or decreases according to their applications. Multi-point sensor is generally used to transmit light for a long distance, and due to its high sensitivity and resolution multi-point sensor is used in chemical solution for measuring different chemical compounds/composition, oil and gas pipeline to measure both the parameter temperature and pressure [11].

4.3 Distributed Sensor

Distributed sensors work on the basis of principle of light scattering like Raman, Rayleigh and Brillouin scattering to measure physical parameters like temperature, intensity of light, vibration and strain and also detect environment change along when light is transmitted inside the fibre [2, 12].

5 Applications of Optical Fibre

Due to unique properties like interference of electromagnetic wave and high immunity, optical fibre is used in sensor technology.

5.1 Temperature Sensor

5.1.1 Fibre Bragg Grating and Distributed Sensor

Optical fibre Bragg grating can sense temperature by specific reflection of wavelength. When temperature changes reflecting wavelength shifts with temperature and due to this it can measure temperature over a wide range.

Distributed optical fibre can sense temperature on the principle of scattering of light. Distributed optical fibre measures temperature in the whole length of optical fibre. This technique is helpful for monitoring large power cable-like structure [13].

5.1.2 Optical Fibre-Based Thermometer

Optical fibre-based thermometer senses temperature through materials which are wrapped or embedded on the optical fibre. Some materials have different properties like emission and absorption which are based on temperature. These types of sensors are used to sense environmental weather condition where traditional electronic sensors are not able to sense.

5.1.3 Wave Sensor

Light passes through optical fibre by reflection or scattering, these types of sensors also measure gas or chemical which is present outside the fibre. Due to the interaction between the two evanescent and light waves, which is spread outside the optical fibre [14].

5.1.4 Pressure Sensor

Micro-bending optical fibre pressure sensor detects pressure micro-bends or very small pressure which comes on the optical fibre during transmission or transfer of data. These sensors detect the pressure underwater like sea or harsh weather conditions also.

5.1.5 Acoustic and Vibration Optical Fibre Sensor

This sensor is used to sense small vibration and acoustic waves or signals. Distributed optical fibre is used to detect the acoustic and interferometric optical fibre sensor measure vibrations, this sensor is used for health monitoring (Fig. 4).

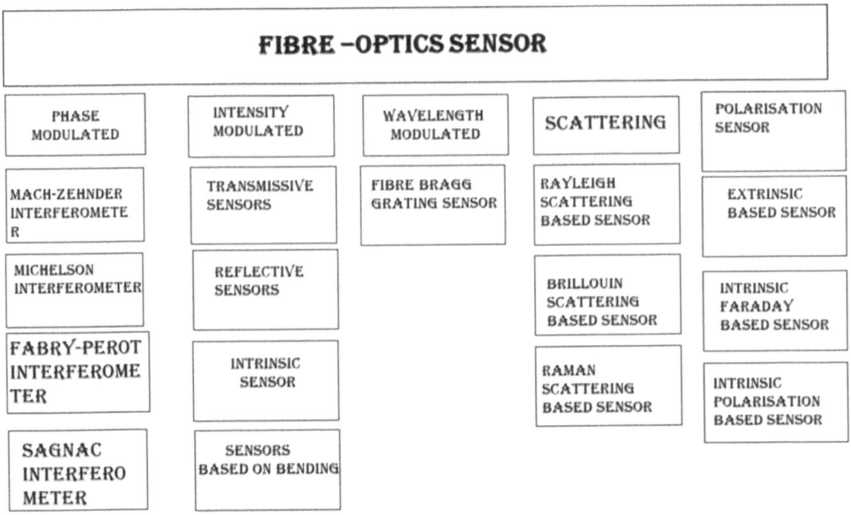

Fig. 4 Different types of optical fibre sensor and application

6 Advancement in Optical Fibre Sensor Using Machine Learning

Over the past few years, optical fibre technology grows significantly and gives engineering benefits over free space and electrical sensor. Generally, conventional sensor measures single physical parameter such as temperature which can be measured by changing the resistance output with respect to thermistor [15]. However, number of parameters recorded by inbuilt arrays of a device (multiple oxide gas sensor). The increased number of sensors in a device means large numbers of data in a single device. Traditional and conventional sensors are not suitable for systematic processing, labelling and analysing the large amount of information [16].

Recent advancement in machine learning algorithms and concepts provides new opportunities and advancement in optical fibre sensor technology. This is very helpful in real-world application process such as healthcare analytic, computer science, economics and analytics [17]. Using machine learning technology, it enhances network efficiency, error detection, improving fault, etc. There are some driven technologies which enhance the future network and security [18] (Fig. 5).

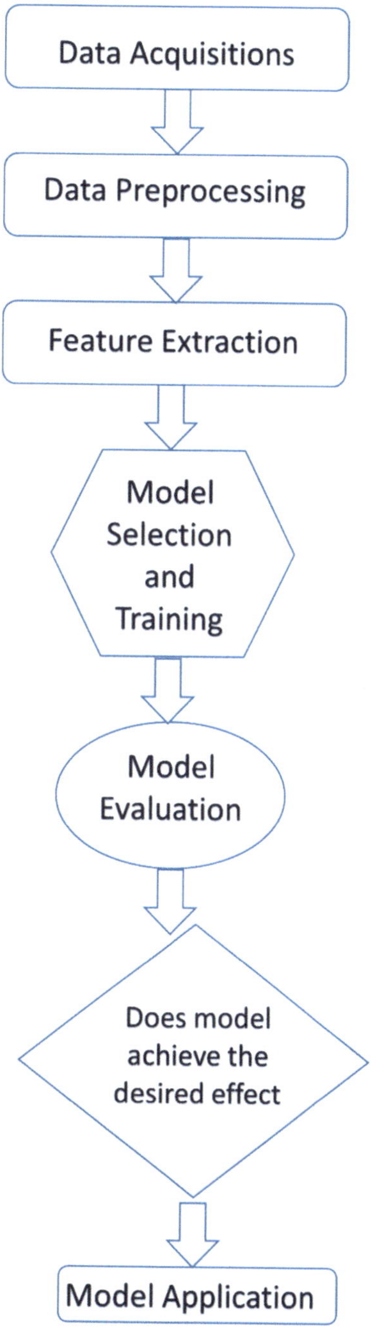

Fig. 5 Machine learning-based spectrum flowchart

6.1 Error Detection and Confinement

6.1.1 Optical Fibre Default Localization

Machine learning algorithms are evaluated using optical time domain reflectometry (OTDRs). OTDRs generally used in fibre to the home network, it stands for broadband Internet connection. Main characteristics of OTDRs are maintenance and monitoring of optical fibre network to pin point the accurate error location in the optical fibre location which reduces the time for repairing and optical return loss [19].

6.1.2 Predictive Health Analysis

Machine learning technology analysis large number of network data and after that it predicts future of a person and informs to a person to address their issue before their impact.

6.1.3 Simulation and Design

Machine learning algorithms have aided to simulate and optimize optical fibre networks, efficiency, cost, stability and also optimize the components of optical fibre like amplifiers and multiplexes and selection of design.

6.1.4 Improvement in Security Network

Machine learning technology examines anomalies in optical fibre network which can easily detect security threat, cybersecurity and taking care of data while transmitting data through optical fibre and due to this sensitive information remain secure.

6.1.5 Self-Healing Network

Machine learning technology being able to develop optical fibre network which can detect error in network and repair itself by adjusting physical parameters to maintain network without interference of human intervention.

7 Advancement in Optical Fibre Sensor Using Deep Learning

Deep learning technology used neural network models for managing the nonlinear and complex system in optical fibre networking [12]. Using these algorithms we can reduce signal distortion, impairment in nonlinear system during transmission which leads to good signal quality and higher rate of data transfer in large amount. Neural network reduces noise and helps in reconstruction of signal using autoencoder, which improves overall efficiency of signal [20–22].

7.1 Signal Processing

During the transfer of data from optical fibre networking different type of noise and nonlinearities occurs due to this signal strength and quality decreases. Deep learning technology helps to identify all the physical parameters in which quality of the signal depend. Recurrent and convolutional neural network helps to reduce nonlinear noise from signals. In this method, there is complicated relationship between the input and output signals which reverse the nonlinearity effect.

7.2 Optimization of Traffic Networking Prediction

Long short-term memory storage and recurrent neural network both can easily analyse the pattern of traffic pattern. Both the models study for the previous traffic data and give future optical traffic prediction.

7.3 Improvement in Optical Fibre Signal

Optical fibre signal improved high output with low latency, minimum loss, etc., which are very difficult to maintain the quality of the optical fibre during video streaming and cloud service.

7.4 Security of Data

Using machine learning algorithms help to protect our data by detecting potential intrusion and malicious activities. Conventional neural network analyse optical traffic pattern to detect suspicious threat during the transfer of data.

7.5 Channel Equalization

Deep learning technology is used to solve complex dynamic of optical fibre channel; it also detects signal which causes to fault or noise during data transmission.

8 Machine Learning Algorithms Used in Optical Fibre Sensor

Machine learning algorithms have numerous applications in the field of optical fibre sensor to improve the quality of the signal, reduce noise or distortion, high accuracy, data prediction, etc. Using machine learning algorithms it helps to solve complex dynamics, environmental sensor, monitoring in health, etc. There are some machine learning algorithms used for optical fibre sensor [22, 23].

(a) **Artificial Neural Network**: This algorithm is used for the application in division of wavelength, intensity of light, measurement of physical parameters in nonlinear devices and learning of complex patterns.
(b) **Random Forest**: This algorithm is used for regression and classification in optical fibre sensor. It is helpful for handling the large dataset, reducing the overlapping of signals and less error. It evaluates the pressure and temperature on the output of the signal.
(c) **Convolution Neural Network**: Conventional neural network is helpful for spatial or grid-like structure. Convolution neural network helps in detection and classification of vehicle or footstep movements. This is generally used for in spatial data, learning of hierarchical feature and recognition.
(d) **Reinforcement Learning**: This algorithm has optimized learning program for dynamic conditions. Reinforcement learning is used in adaptive optical fibre sensor. It measures optimal measurements for dynamic environment conditions.
(e) **Analysis of Time Series**: These algorithms play a key role in smart sensor. It is used for monitoring and foresight parameters which changes over time. This algorithm is used for specific applications like pressure, temperature and strain.
(f) **Gaussian Process Regression**: Gaussian process regression technique is used for prediction of temperature and strain which are changed over time. This algorithm deals with distorted data and gives probability predication and gives effective database.
(g) **Clustering Algorithms**: Clustering algorithms are used to detect same group pattern. It is helpful for data analysis in strain events or temperature.
(h) **Analysis of Principle Components**: This technique is used to reduce computation complexity, eliminate signal distortion and store important information during sensing the parameter.
(i) **K-Nearest Neighbours**: These algorithms are used for small database easy for implantation. But it senses some specific parameters like distance metrics.

8.1 *Limitation and Opportunities of Machine Learning and Artificial Intelligence*

Artificial intelligence and machine learning both have various applications but there are some limitations in both the techniques.

8.1.1 Data Availability

Machine learning and artificial intelligence techniques depend upon the availability of the data. If available data have error or inaccurate, we are not available design model because both the techniques depend upon the large dataset. Collection of large datasets may also concern about the security of the data, privacy, regulation, etc. [23, 24].

Main challenge of the design model using this technique is collection of accurate data or labelled dataset which depends upon interest of the events. Audio version and computer version have been examined on large dataset. However, most of the does not shared for public/open. Possession of distributed optical fibre sensor is expensive [4].

8.1.2 Big Data

Distributed optical fibre sensor produces large amount of dataset because of large sensing area and small spatial resolution. It contains system scale prohibitively with sensing range. So, we required to generate labelled dataset for designing model and compress algorithm for raw sensor. But Brillouin-based distributed optical fibre sensors are more accurate for small frequency range for large data issue.

8.1.3 Noise from Source

Low signal-to-noise ratio is one of the major afflictions for several optical fibre technologies. There are several types of noise like polarization, amplified spontaneous emission, relative intensity noise and thermal shot noise. These noises degrade the quality of the signal. Intensity fading is also observed in Rayleigh backscattered traces. Polarization fading is the major problem in heterodyne detection.

8.1.4 Cross-Sensitivity Challenges

Cross-sensitive challenge occurs in the sensitive model (environment or specific data) or modification in the input data. For a specific dataset for particulars application is useful in different domain. Efficiency of model significantly drops due to domain shift.

In case of deep neural network are highly sensitive precisely crafted in the perturbated or underrepresented input due to this predication given by model is inaccurate with high assurance.

8.1.5 Nonlinear Dynamic Sensor Data

Nonlinear dynamic means that small variation or error in the input dataset leads to unpredictable output or, in other words, we can say that output of our design model is not dependent on the input. Output of the sensor depends upon the behaviour of the system.

9 Opportunities of Machine Learning and Artificial Intelligence

9.1 Operational Efficiency and Automation

Artificial intelligence and machine learning technology both optimize the large dataset from the industries, management for customer service and creative task, which leads to cost-effective and improved quality of the product [25].

9.2 Improvement of Predicative Analytic

Artificial intelligence optimizes the large number of datasets to study the pattern and trends which helps in marketing and healthcare which leads to high-quality outcomes and predicative analysis.

9.3 Creative Application in Services and Product

Both the techniques help to generate new music, design, art and create new content which assist human. These techniques also help to generate new human creators. Artificial intelligence also enables to produce new product, services and enterprise which were have previously happened like smart home device, healthcare, etc. [24].

9.4 Advancement in Medicine and Treatment

Artificial intelligence studies the genetic information and predicts treatment plan for an individual person. By studying the large datasets using machine learning and artificial intelligence, early diagnosis of the diseases and planning the treatment are possible for human which lead to better outcome.

9.5 Solution of Complex Problem

Artificial intelligence can easily solve the complex problem, which is difficult to solve by human-like climate and protein modelling. Artificial intelligence also eases to solve climate science, genomics, etc.

9.6 Improvement in Skill Adaptive

Artificial intelligence helps in simulation for training in the field of medicine, military operation, etc.

9.7 Assistive Language Translation

Artificial intelligence enables new techniques for person who have disabilities like hearing, speech impaired with computer vision.

10 Conclusion and Future Perspective

This chapter provides a compressive study of development of intelligent and smart study using optical fibre sensor. For high accuracy and enhancing the efficiency of optical fibre sensor, it is combined with artificial intelligence which helps for the development of smart society and gives accurate prediction. From research and development of optical fibre sensor technology, infinite benefits like development of multiparameter sensing capabilities, pattern study by quasi-distribution and generation of large labelled data are provided. Artificial intelligence algorithms help in health monitoring disabled person, machine translation, and also help in solving complex problem which humans are not able to solve.

Acknowledgements Authors would like to thank vice-chancellor of Bharati Vidyapeeth University and Netaji Subhas University of Technology for providing the financial support.

References

1. R.K. Gangwar, A.K. Pathak, F. Chiavaioli, M.H. Abu Bakar, Y.M. Kamil, M.A. Mahdi, V.K. Singh, Optical fiber SERS sensors: Unveiling advances, challenges, and applications in a miniaturized technology. Coord. Chem. Rev. **510**(2024). https://doi.org/10.1016/j.ccr.2024.215861
2. S. Addanki, I.S. Amiri, P. Yupapin, Review of optical fibers-introduction and applications in fiber lasers. Results in Physics **10**, 743–750 (2018). https://doi.org/10.1016/j.rinp.2018.07.028
3. I.H. Sarker, Machine learning: algorithms, real-world applications and research directions. SN Computer Science **2**(3), 160 (2021). https://doi.org/10.1007/s42979-021-00592-x
4. M. Elsherif, A.E. Salih, M.G. Muñoz, F. Alam, B. AlQattan, D.S. Antonysamy, ...H. Butt, Optical fiber sensors: working principle, applications, and limitations. Adv. Photon. Res. **3**(11) (2022). https://doi.org/10.1002/adpr.202100371
5. N. Lalam, S. Bukka, H. Bhatta, M. Buric, P. Ohodnicki, R. Wright, Achieving precise multi-parameter measurements with distributed optical fiber sensor using wavelength diversity and deep neural networks. Commun. Eng. **3**(1), 121 (2024). https://doi.org/10.1038/s44172-024-00274-5
6. K.S.C. Kuang, S.T. Quek, M. Maalej, Assessment of an extrinsic polymer-based optical fibre sensor for structural health monitoring. Meas. Sci. Technol. **15**(10), 2133 (2004). https://doi.org/10.1088/0957-0233/15/10/024
7. G.A. Lashari, F. Mumtaz, Z. Ai, Y. Dai, Recent advancements and future challenges in hybrid optical fiber interferometers. Optik **282**, 170860 (2023). https://doi.org/10.1016/j.ijleo.2023.170860
8. K. Bremer, T. Reinsch, G. Leen, B. Roth, S. Lochmann, E. Lewis, Pressure, temperature and refractive index determination of fluids using a single fibre optic point sensor. Sens. Actuators, A **256**, 84–88 (2017). https://doi.org/10.1016/j.sna.2017.01.025
9. L. Fazzi, S. Valvano, A. Alaimo, R.M. Groves, A simultaneous dual-parameter optical fibre single sensor embedded in a glass fibre/epoxy composite. Compos. Struct. **270**(2021). https://doi.org/10.1016/j.compstruct.2021.114087
10. A. Ghatak, K. Thyagarajan, Single-mode optical fiber sensors, in *Introduction to fiber optics* (pp. 397–410) (1998)
11. X. Liu, X. Zhang, Y. Liu, Z. Liu, W. Peng, Multi-point fiber-optic refractive index sensor by using coreless fibers. Optics Commun. **365**, 168–172 (2016). https://doi.org/10.1016/j.optcom.2015.10.071
12. Y. LeCun, Y. Bengio, G. Hinton, Deep learning. Nature **521**(7553), 436–444 (2015). https://doi.org/10.1038/nature14539
13. R.K. Chourasia, A. Katti, Introduction to bragg fiber, in R.K. Chourasia, A. Katti (Eds.), *Bragg Fibers: From Optical Properties to Applications*, pp. 1–16. Cham: Springer Nature Switzerland (2024)
14. X. Wang, W. Zhao, R. Ma, J. Zhuo, Y. Zeng, P. Wu, J. Chu, A novel high accuracy fast gas detection algorithm based on multi-task learning. Measurement **228**, 114383 (2024). https://doi.org/10.1016/j.measurement.2024.114383
15. A. Venketeswaran, N. Lalam, J. Wuenschell, P.R. Ohodnicki, M. Badar, K.P. Chen, ...M. Buric, Recent advances in machine learning for fiber optic sensor applications. Adv. Intell. Syst. **4**(1) (2021). https://doi.org/10.1002/aisy.202100067
16. Y. Liu, X. Tan, Y. Bao, Machine learning-assisted intelligent interpretation of distributed fiber optic sensor data for automated monitoring of pipeline corrosion. Measurement **226**(2024). https://doi.org/10.1016/j.measurement.2024.114190
17. Y. Zhou, Y.-N. Zhang, Q. Yu, L. Ren, Q. Liu, Y. Zhao, Application of machine learning in optical fiber sensors. Measurement **228**(2024). https://doi.org/10.1016/j.measurement.2024.114391
18. N. Ha, K. Xu, G. Ren, A. Mitchell, J.Z. Ou, Machine learning-enabled smart sensor systems. Adv. Intell. Syst. **2**(9) (2020). https://doi.org/10.1002/aisy.202000063

19. E. Reyes-Vera, A. Valencia-Arias, V. García-Pineda, E.F. Aurora-Vigo, H. Alvarez Vásquez, G. Sánchez, Machine learning applications in optical fiber sensing: a research agenda. Sensors **24**(7), 2200 (2024). Retrieved from https://www.mdpi.com/1424-8220/24/7/2200
20. B. Rahmani, D. Loterie, G. Konstantinou, D. Psaltis, C. Moser, Multimode optical fiber transmission with a deep learning network. Light Sci. Appl. **7**(1), 69 (2018). https://doi.org/10.1038/s41377-018-0074-1
21. B. Karanov, M. Chagnon, V. Aref, D. Lavery, P. Bayvel, L. Schmalen, *Optical Fiber Communication Systems Based on End-to-End Deep Learning : (Invited Paper)*. Paper presented at the 2020 IEEE Photonics Conference (IPC) (2020)
22. F.N. Khan, Q. Fan, C. Lu, A.P.T. Lau, Chapter 21—Machine learning methods for optical communication systems and networks, in *Optical Fiber Telecommunications VII*. ed. by A.E. Willner (Academic Press, 2020), pp.921–978
23. L.K. Oxenløwe, F. Klejs, M. Lillieholm, P. Guan, F.D. Ros, P.M. Kaminski, M. Galili, Chapter 6—Optical processing and manipulation of wavelength division multiplexed signals, in *Optical Fiber Telecommunications VII*. ed. by A.E. Willner (Academic Press, 2020), pp.233–299
24. X. Pan, X. Wang, B. Tian, C. Wang, H. Zhang, M. Guizani, Machine-learning-aided optical fiber communication system. IEEE Network **35**(4), 136–142 (2021). https://doi.org/10.1109/MNET.011.2000676
25. A. Sarirete, Z. Balfagih, T. Brahimi, M.D. Lytras, A. Visvizi, Artificial intelligence and machine learning research: towards digital transformation at a global scale. J. Ambient. Intell. Humaniz. Comput. **13**(7), 3319–3321 (2022). https://doi.org/10.1007/s12652-021-03168-y

The Role of Artificial Intelligence in Enhancing Optical Fiber Sensors

Shreya, Satyam Rawat, Bhawana Chand, Peeyush Phogat, Ranjana Jha, and Sukhvir Singh

Abstract The integration of artificial intelligence (AI) with optical fiber sensors (OFSs) represents a paradigm shift in sensing technology, offering unprecedented improvements in sensitivity, accuracy, and data processing capabilities. This chapter delves into the synergistic relationship between AI and OFSs, highlighting the transformative impact AI algorithms have on enhancing the performance and functionality of these sensors. Numerous artificial intelligence (AI) methods, including machine learning, deep learning, and neural networks, have been investigated along with their uses in improving signal processing, sensor design, and real-time data analysis. There are case studies that show how AI-enhanced OFSs have been successfully applied in a variety of industries, such as industrial automation, healthcare, environmental monitoring, and structural health monitoring. The chapter also addresses the challenges and prospects of this interdisciplinary approach, emphasizing the potential for AI to drive the next generation of intelligent sensing systems. By leveraging the strengths of both AI and optical fiber technologies, the aim is to provide insights into how this fusion can lead to more efficient, robust, and adaptable sensing solutions.

Keywords AI · Optical fiber sensors · Machine learning · Deep learning · Signal processing · Sensor optimization · Intelligent sensing systems

Shreya (✉) · S. Rawat · B. Chand · P. Phogat · R. Jha
Research Lab for Energy Systems, Department of Physics, Netaji Subhas University of Technology, Dwarka, New Delhi, India
e-mail: shreyasharma.aug15@gmail.com

Shreya · P. Phogat
CSIR-National Institute of Science Communication and Policy Research, New Delhi, India

S. Singh
Department of Nanotechnology, Delhi Skill Entrepreneurship University, New Delhi, India

V. Bhardwaj et al. (eds.), *Optical Fiber Sensors and AI*, Progress in Optical Science and Photonics 34, https://doi.org/10.1007/978-981-96-6158-9_3

1 Introduction

Due to their high sensitivity, dependability, and versatility, optical fiber sensors (OFSs) have emerged as a key component in numerous applications because of the rapid evolution of technology. At the same time, major advancements in the field of artificial intelligence (AI) have transformed the processing and interpretation of data. The convergence of these two powerful technologies—optical fiber sensors and AI—holds immense potential for creating sophisticated, intelligent sensing systems capable of addressing complex real-world challenges [29]. In this chapter, the dynamic synergy between optical fiber sensors and AI is explored, focusing on how AI can enhance the performance and functionality of these sensors.

1.1 Overview of Optical Fiber Sensors

OFSs are devices that use optical fibers—thin, flexible strands of glass or plastic as a medium to detect and measure various physical, chemical, or biological parameters. Because of their special benefits and capacities, OFSs, as seen in Fig. 1a, have become a crucial technology in many different industries. Researchers started investigating the use of optical fibers for sensing applications in the 1970s, which led to the invention of optical fiber sensors. The initial focus was on telecommunications, but it quickly expanded to include industrial, medical, and environmental applications. The versatility of OFSs stems from their intrinsic properties, including high sensitivity, ability to operate in harsh environments, and the immunity to electromagnetic interference. OFSs have revolutionized the field of sensing technology, providing unparalleled advantages in terms of sensitivity, accuracy, and versatility [30].

The basic principle of operation of OFSs as shown in Fig. 1b involves the modulation of light properties such as intensity, phase, polarization, or wavelength in response to external stimuli. The changes in these properties are then analyzed to infer the measurements of interest. The transducer alters a specific parameter of the optical system, such as intensity, wavelength, polarization, or phase, resulting in a modification of the optical signal's characteristics that the detector receives. The transmission of signal works on the principle of total internal reflection in which light gets confined to the core by reflecting at higher index-clad layers, making sure that almost no light leaks out over long distances [4].

Dissemination mechanisms are the main modulation method applied to optical fiber sensors. It captures the changes in light intensity due to different factors like bending, displacement, or other chemical absorption. This results in variations of light intensity, which is detected and is therefore correlated with the corresponding external parameter. Phase modulation is the term used to describe changes that take place in the light wave's phase. The modulated light signals are converted into electrical signals using photodetectors, which are then processed and analyzed to extract

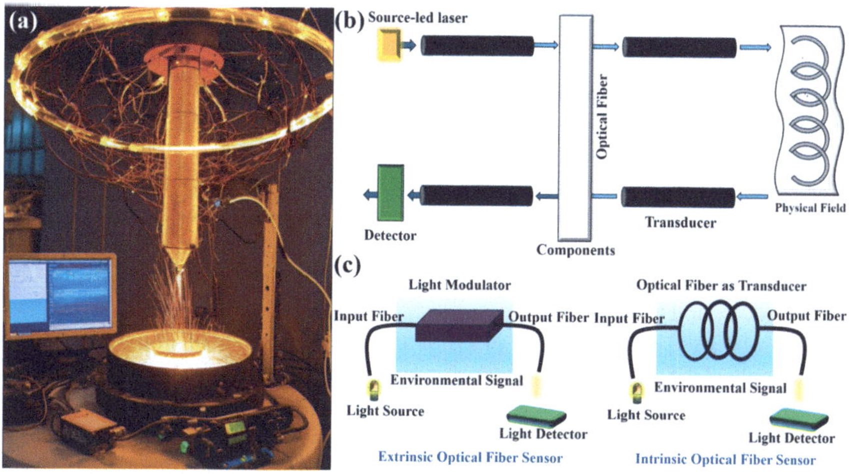

Fig. 1 **a** Optical fiber sensor. This picture is AI generated using lexica. art, **b** schematic diagram showing principle of an optical fiber, and **c** intrinsic- and extrinsic-type fiber optics

meaningful data about the external environment. Advanced signal processing techniques, including digital signal processing (DSP) [31] and AI algorithms [13], are employed to enhance the sensitivity, accuracy, and real-time processing capabilities of these sensors.

Optical fiber sensors are categorized into two main types: intrinsic and extrinsic as depicted in Fig. 1c. Intrinsic sensors have the sensing mechanism within the fiber itself, while extrinsic sensors use the fiber to transmit light to and from an external sensing element [2]. Numerous types of sensors fall within these categories, including distributed fiber optic sensors like Distributed Acoustic Sensing (DAS) and Distributed Temperature Sensing (DTS), intensity-based sensors, interferometric sensors, and Fiber Bragg Grating (FBG) sensors [28].

Despite their advantages, OFSs face challenges such as sensitivity to environmental factors, high costs, and complexity in integration and deployment. Addressing these challenges requires ongoing research and innovation, particularly in enhancing signal processing techniques and developing cost-effective manufacturing processes. Addressing these challenges requires ongoing research and innovation, particularly in enhancing signal processing techniques and developing cost-effective manufacturing processes [16].

1.2 *Emergence of AI in Sensing Technologies*

There are many different types of sensors in these categories. Artificial Intelligence (AI) has transformed several sectors by allowing robots to learn from data, spot patterns, and make judgments with little assistance from humans. Because AI provides advanced capabilities for analyzing and understanding the massive volumes of data provided by sensors, the field of sensing technologies has profited greatly from it. An important turning point in the development of intelligent systems has been reached with the combination of AI and sensing technologies, such as OFSs (AI for OFS, 2023). The early 2000s saw a surge in the use of AI in sensing technologies due to improvements in machine learning algorithms, more processing power, and the accessibility of big datasets. Artificial intelligence (AI) methods, like machine learning and deep learning, are crucial for improving sensor performance because of their remarkable performance in tasks like classification, regression, clustering, and anomaly detection [8].

AI can be used in the context of OFSs to solve a number of important issues. For example, OFSs frequently produce noisy, complex data from which relevant information must be extracted using sophisticated processing methods. Artificial intelligence (AI) algorithms can efficiently reduce noise, increase signal quality, and raise measurement accuracy. AI can also facilitate real-time data processing, which enables prompt identification and reaction to modifications in the monitored environment [1]. The integration of AI with OFSs extends beyond signal processing to include intelligent decision-making systems. These systems can autonomously monitor and interpret sensor data, providing actionable insights and triggering automated responses. For example, in structural health monitoring, an AI-powered system can continuously assess the condition of a structure and alert maintenance teams of potential issues before they become critical [7].

The advent of artificial intelligence (AI) in sensing technologies is a revolutionary advancement that could greatly improve OFS capabilities. By harnessing the power of AI, we can overcome existing limitations, unlock new functionalities, and drive innovation across various applications. The continued evolution of AI and its integration with sensing technologies will undoubtedly shape the future of intelligent systems.

2 AI Techniques for Improved OFSs

The rapid evolution of AI has catalyzed significant advancements across various technological domains, including sensor technology. AI techniques, encompassing machine learning (ML), deep learning (DL), and neural networks (NNs), have introduced sophisticated methods for data processing, analysis, and interpretation. These techniques have proven to be particularly transformative in the realm of OFSs, where the sheer volume and complexity of data generated necessitate advanced analytical tools for meaningful insights [7].

2.1 Overview of AI Techniques

AI techniques, particularly ML, DL, and NNs, play a vital role in enhancing the functionality and performance of OFSs which are discussed in this section.

Without explicit programming, machines can learn and get better from experience thanks to machine learning, a subfield of artificial intelligence. It utilizes algorithms capable of analyzing and interpreting intricate data patterns, making it particularly suitable for enhancing sensor data analysis. ML techniques depicted in Fig. 2 such as regression, classification, and clustering are widely used in processing and interpreting sensor data. It entails developing algorithms that enable systems to generate predictions and learn from data. Labeled data is used in supervised learning to train models that categorize or forecast data elements [1].

DL is a subfield of machine learning that uses multi-layered neural networks (NNs) to model intricate correlations and patterns in data. For high-dimensional data, like the signals generated by optical fiber sensors, DL works especially well. Techniques such as Convolutional NNs (CNNs) and Recurrent NNs (RNNs) offer powerful tools for feature extraction, pattern recognition, and temporal data analysis. CNNs are particularly useful for spatial data analysis, such as image processing, and are defined by layers of convolutions. RNNs, including Long Short-Term Memory (LSTM) networks, are effective for sequential data, capturing temporal dependencies [10].

NNs are the foundation of DL and consist of interconnected nodes (neurons) that work in unison to process data resembling human neurons depicted in Fig. 3a. They are capable of learning nonlinear relationships and are highly adaptive, making them suitable for various sensor applications. Neural networks can be designed and trained to perform specific tasks such as anomaly detection, predictive maintenance,

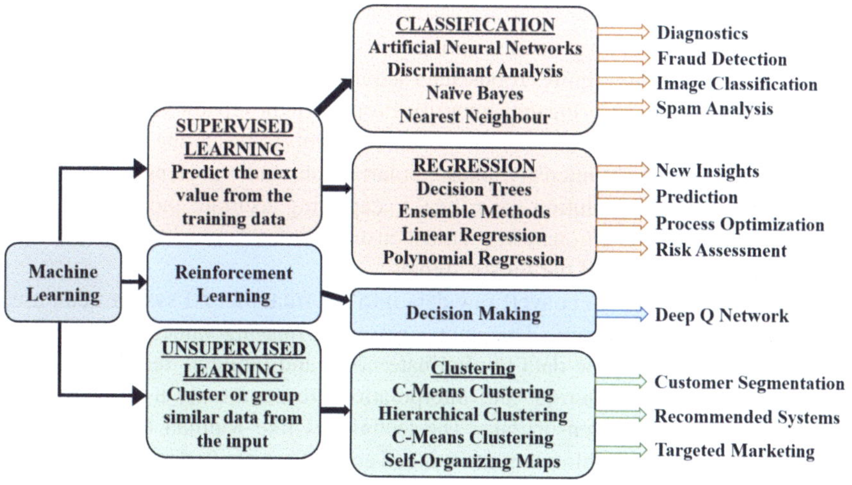

Fig. 2 Grouping machine learning approaches into several subcategories

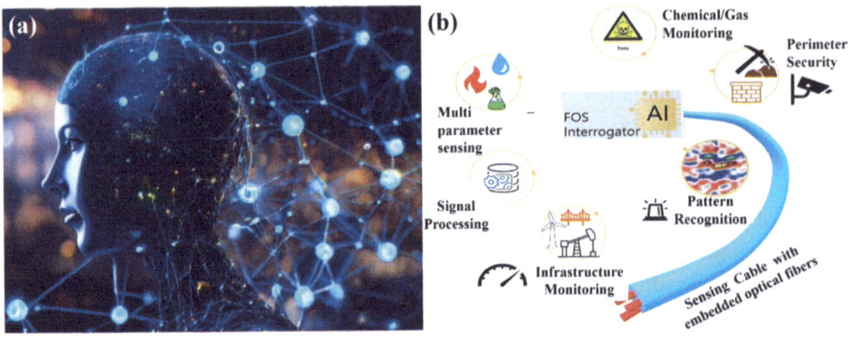

Fig. 3 **a** Neural networks in a human being. This image is AI generated using lexica. art, **b** depicts cables embedded with optical fiber linked to an advanced FOS integrator using AI. Reproduced from Ref. [32]. CC BY License. Published by Wiley–VCH GmbH

and real-time monitoring. An input layer, one or more hidden layers, and an output layer make up the fundamental structure. By introducing nonlinearities, activation functions like sigmoid, tanh, or ReLU (rectified linear unit) allow NNs to simulate intricate relationships in data [1].

2.2 *Data Processing for OFSs*

Effective AI model performance relies on high-quality data acquisition and preprocessing, particularly for OFSs that generate vast amounts of raw data. The first step in enhancing optical fiber sensors with AI is the effective acquisition and preprocessing of data. This involves collecting raw sensor data and preparing it for analysis.

a. Data acquisition: OFSs generate continuous streams of data that need to be accurately captured. This requires robust data acquisition systems capable of handling high data rates and ensuring data integrity. Proper synchronization and calibration of sensors are crucial to obtain reliable data. It involves collecting raw data from OFSs, including light intensity, phase, polarization, and wavelength shifts. The sampling rate and resolution are critical to capturing accurate and detailed data. Advanced data acquisition systems must ensure minimal noise and distortion to preserve the integrity of the sensor signals.
b. Data preprocessing: To convert raw data into a format that AI systems can use, preprocessing is necessary. This entails resolving missing values, leveling data scales, and cleaning the data to eliminate noise and outliers. Techniques such as data smoothing, filtering, and interpolation are commonly used in preprocessing optical fiber sensor data. The techniques like Kalman filters, Fourier transforms, and smoothing (e.g., moving averages) are applied to mitigate noise and enhance signal quality. **Normalization and scaling** ensure uniformity in data using methods such as min–max scaling represented as Eq. 1. In cases of

limited data, synthetic data generation (data augmentation) can create additional samples to improve model robustness [3].

$$x\prime = \frac{x - \min(x)}{\max(x) - \min(x)} \quad (1)$$

c. Feature extraction and selection: AI model effectiveness is enhanced by extracting and selecting relevant features from sensor data. The process of feature extraction entails turning unprocessed sensor data into a collection of useful characteristics. For OFSs, features could include amplitude variations, frequency components, phase shifts, and temporal patterns. To extract these features, sophisticated signal processing methods like Fourier analysis and wavelet transforms are frequently used.

Using statistical techniques (ANOVA and chi-square tests), regularization techniques (LASSO, which adds a penalty to the regression coefficients), and recursive feature elimination (RFE), which iteratively eliminates the least significant features, feature selection entails identifying the most pertinent features [37].

2.3 *Implementation of AI Models for Sensor Data Analysis*

Applying AI models to analyze sensor data involves selecting appropriate algorithms and training them on the processed data.

a. **Supervised Learning Models**: These include classification models (decision trees, support vector machines) and regression models (linear, polynomial) that are trained on labeled data to provide predictions or categorize incoming data points. Using linear relationships between input features and target variables, the linear regression model forecasts continuous outcomes. To classify data, support vector machines (SVM) identify the best hyperplane to divide data points into distinct classes. An ensemble of decision trees called a random forest increases the accuracy of predictions.
b. **Unsupervised Learning Models**: These models, like k-means clustering and hierarchical clustering, are used to find patterns or groupings in unlabeled data, making them suitable for anomaly detection and exploratory data analysis. K-means clustering groups data into clusters based on feature similarities. Principal component analysis (PCA) reduces data dimensionality while preserving variance facilitating visualization and further analysis.
c. **Advanced AI Models**: CNNs are effective for spatial data analysis, while RNNs and LSTM networks excel at temporal data analysis, making them ideal for time-series sensor data.

Implementing AI models for optical fiber sensor data involves several steps, from model selection and training to optimization and validation.

a. **Model Training**: To reduce prediction error, the model is fed training data and parameters are changed during the training process. For optimization, methods like gradient descent and backpropagation (for neural networks) are frequently employed.
b. **Hyperparameter Tuning**: Model performance is greatly impacted by hyperparameters (such as learning rate and number of layers). The best hyperparameter settings can be found with the aid of strategies like grid search and random search.
c. **Model Validation**: Model performance and generalizability are evaluated using cross-validation methods like k-fold cross validation. To guarantee robustness, this entails dividing the data into training and validation sets and repeating the training procedure.
d. **Performance Metrics**: Quantitative assessments of model performance are provided by metrics including accuracy, precision, recall, F1-score, and mean squared error (MSE), which direct future improvements [15].

2.4 AI-Driven Signal Processing in OFSs

The integration of AI in signal processing for OFSs offers advanced methodologies to mitigate noise, enhance signal quality, recognize patterns, and facilitate real-time monitoring and decision-making. This chapter delves into the scientific principles and methodologies underpinning AI-driven signal processing in optical fiber sensors [11].

a. **Noise Reduction and Signal Enhancement**: Interference in fiber optic sensors arises from environmental factors, electronic components, and the sensors themselves. Traditional noise suppression techniques, such as filtering, often fall short in preserving signal integrity. However, ML and DL offer advanced methods for noise reduction and signal enhancement.

Machine learning algorithms like SVM, k-Nearest Neighbors (k-NN), and Random Forests can be trained on historical data to identify and filter out noise. These algorithms detect statistical features of both noise and signal, applying adaptive filters to improve signal quality. For instance, SVM can distinguish between noise and signal based on their features, thereby minimizing noise while preserving the signal [26].

More advanced algorithms, such as CNN [14] and RNN [5], provide superior capabilities for signal enhancement. CNNs are particularly effective in extracting spatial relations and enhancing optical sensor signal patterns. RNNs, designed for time-series data, utilize temporal dependencies and correlations to improve signal quality.

Additionally, the wavelet transforms, a powerful signal processing technique, can be combined with AI to further enhance noise reduction and signal quality. Wavelet transform decomposes the signal into various frequency components, isolating noise effectively. AI models can optimize the parameters of the wavelet transform, ensuring optimal signal reconstruction and enhancement.

b. **Pattern Recognition and Classification**: Optical fiber sensors produce big data with much environmental information embedded in it. Some of these techniques include neural networks, decision trees, and ensemble methods that aid in figuring out patterns and categorization of these anomalies from the data. These models are trained on labeled data to define the normal and abnormal conditions. Algorithms like k-means, PCA [27] for dimensionality reduction, and autoencoder of unsupervised learning help us to detect the abnormalities in the unseen data. Thus, raw outliers are filtered out at the unsupervised model level, and at the level of the supervised models their refinement is carried out, thereby increasing the reliability of pattern recognition in optical fiber sensor data.

c. **Real-Time Monitoring and Decision Making**: Real-time control and monitoring with AI implies the use of AI models on edge devices in order to have low latency and applying such tools as TensorFlow Lite and Edge AI for the signal processing as depicted in Fig. 3b. AI also allows an organization to conduct predictive maintenance through the use of data that are collected and subsequently analyzed in real time to predict the likelihood of equipment failure. For instance, evaluating machine vibration indicates that maintenance crews have to be informed of possible failures soon. Applying AI in signal processing helps to minimize the degree of expert's interference. In SHM, AI can inform maintenance personnel of the specific situation or trigger response measures for the detected problem. Many IoT [34] systems are connected with optical fiber sensor to further expand the scope of the application, so that it can integrate data analysis functions of various sensors for the purpose of cooperation. opinion 3 indicates that optical fiber sensors can improve safety and plan resources.

3 AI-Enhanced OFSs

AI-enhanced OFSs represent a significant leap in the capabilities of sensing technology. These ML and DL algorithms process and interpret data with greater accuracy and efficiency. Below are more detailed examples showcasing how AI-enhanced OFSs are being utilized across various applications [1].

a. **Distributed Acoustic Sensing (DAS) Systems** use optical fibers to detect acoustic signals along their length, providing continuous monitoring over large areas with high precision. For example, in pipeline monitoring, AI can differentiate between normal operational sounds and those indicative of leaks or tampering. Over time, the system can increase its detection accuracy by learning from past data, which lowers false alarms and allows for prompt response.

b. **Optical Fiber Gyroscopes (FOGs)** as depicted in Fig. 4 are used for precise rotational sensing, critical in navigation systems for aerospace and underwater applications. AI algorithms enhance FOGs by compensating for drift and noise, which are common challenges in gyroscopic measurements. By applying machine learning techniques, these sensors can predict and correct deviations, resulting in more accurate orientation and navigation data.

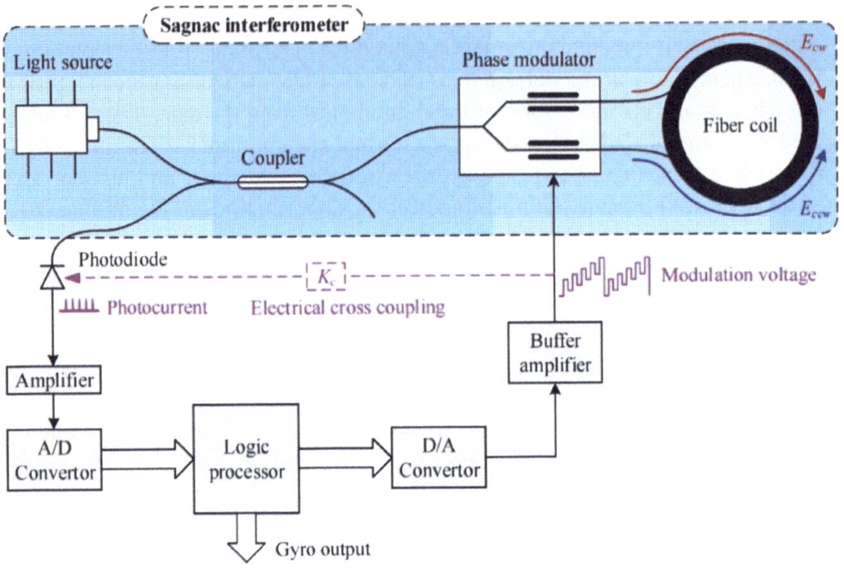

Fig. 4 Schematic of a closed-loop FOG. Reproduced from Ref. [17]. CC BY License. Licensee Optica Publishing Group

c. **Optical Coherence Tomography (OCT)** is a non-invasive imaging method commonly employed in medical diagnostics, particularly in ophthalmology, as illustrated in Fig. 5 (a & b). AI-enhanced OCT systems utilize deep learning models to analyze the high-resolution images produced by OCT scans. These models can identify and classify various retinal conditions, such as macular degeneration and diabetic retinopathy, with greater accuracy than conventional methods. AI algorithms can detect subtle features and patterns in the images, facilitating early diagnosis and treatment planning by training on large datasets.
d. **Fiber Optic Chemical Sensors** detect the presence of specific chemicals by measuring changes in light properties as they interact with the chemical substances. For example, in industrial environments, AI can analyze sensor data to detect trace amounts of hazardous chemicals, even in the presence of background noise.
e. **Distributed Temperature-Sensing (DTS) Systems**, in such systems optical fibers are deployed along extensive lengths to monitor temperature variations with high spatial resolution.
f. **Fiber Bragg Grating (FBG) Sensors** detect changes in the wavelength of light reflected from a Bragg grating etched into the optical fiber to determine temperature and strain. These sensors can anticipate and adjust for noise and nonlinearities, increasing their sensitivity and accuracy, by using deep learning models trained on past sensor data.

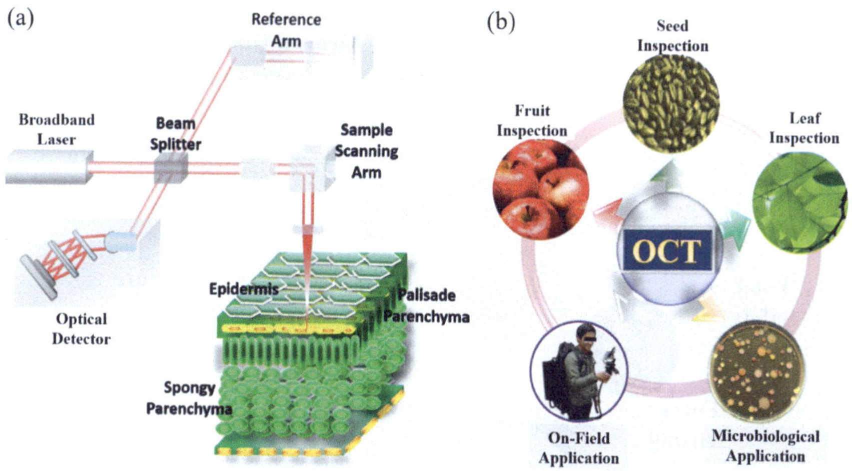

Fig. 5 **a** Optical coherence tomography system schematic. **b** Agricultural OCT applications. Reproduced from Ref. [21].

4 Applications of AI-Based OFSs

AI-based OFSs have revolutionized various fields by enhancing the precision, sensitivity, and functionality of traditional sensing systems. Real-time data processing and analysis are made possible by these sophisticated sensors' integration with AI algorithms, which yield precise and useful insights for a variety of applications. The incorporation of AI algorithms enables these sensors to handle complex datasets, detect anomalies, and optimize performance, making them invaluable in healthcare, environmental monitoring, structural health monitoring, and industrial automation as shown in Fig. 6 [32].

a. **Healthcare: Biomedical Sensing and Diagnostics**:

In healthcare, AI-enhanced OFSs play a vital role in biomedical sensing and diagnostics. As seen in Fig. 7, these sensors are crucial for non-invasive diagnostics and ongoing patient monitoring since they can identify even the smallest physiological changes and provide real-time vital sign monitoring. AI-based OFSs can measure biochemical markers in body fluids with high sensitivity. For instance, fiber optic biosensors can detect glucose levels in diabetic patients by measuring the fluorescence intensity changes in response to glucose concentration [6]. These sensors, integrated with AI algorithms, can analyze complex biological signals to diagnose diseases.

ML models trained on vast datasets can identify patterns associated with specific health conditions, enabling personalized treatment, early diagnosis, and early diagnosis plans. For example, AI algorithms can classify cancerous and non-cancerous tissues based on optical spectroscopy data [12]. Sajan et al. studied silicon photonics

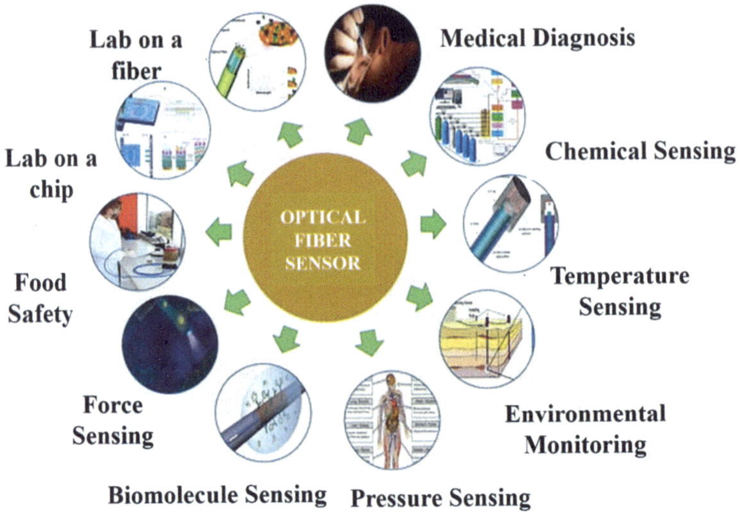

Fig. 6 Various applications of optical fiber sensors

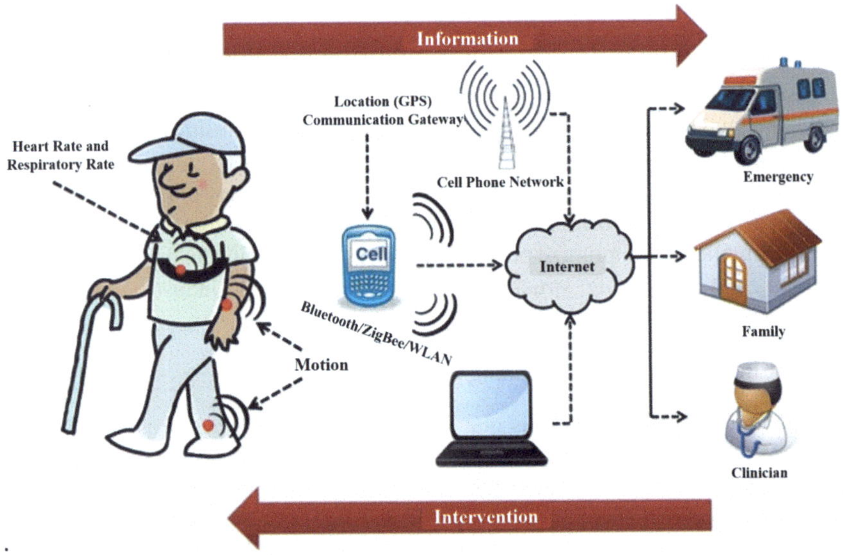

Fig. 7 A wearable sensor-based remote health monitoring device. Reprinted with the permission from Ref. [18]. CC BY Licence. Licensee BioMed Central Ltd

biosensors for cancel cell detection [20]. The research investigates innovative silicon photonics (SiPh) technology for detecting cancer cells by leveraging their optical properties. It also discusses key characteristics of biosensors, including user-friendliness, cost-effectiveness, and sensitivity. Additionally, the paper offers a comparative review of different SiPh architectures, including integrated chips, rings, waveguides, photonic crystals (PhC), and sensor arrays. Another study by Zha et al. investigated the use of stretchable elastomer optical fiber in an intelligent wearable photonic sensing system for remote healthcare monitoring [35]. They successfully used a 1-D convolutional neural network to identify speech-related activities and knee joint actions, as well as to track breathing patterns, heart rate, pulse, facial micro-expressions, and joint movements, with over 90% accuracy in identifying each activity. These benefits point to its potential as a medical instrument and point to promising opportunities for remote health monitoring.

b. **Environmental Monitoring: Air and Water Quality Sensing**

AI-based OFSs are pivotal in environmental monitoring, providing accurate and continuous data on water and air quality as shown in Fig. 8. These sensors help in detecting pollutants and assessing environmental conditions in real time. OFSs can detect gases and particulate matter in the air. AI algorithms process the sensor data to identify pollutant levels and sources. The Beer–Lambert law describes the attenuation of light through a medium, which is fundamental in gas detection [25]. OFSs equipped with AI can monitor various water quality parameters, such as pH, turbidity, and chemical contaminants. These sensors utilize AI algorithms to analyze spectral data, providing real-time insights into water quality and detecting potential contamination events [36].

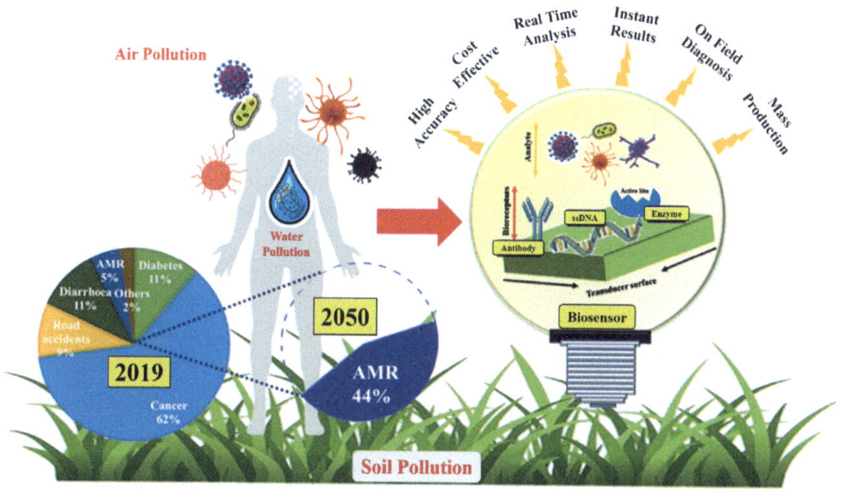

Fig. 8 Optical biosensors for environmental monitoring. Reproduced with the permission from Ref. [24]. Copyright © 2023 Elsevier Inc. All rights reserved

c. **Industrial Automation: Process Control and Optimization**

In industrial automation, AI-based OFSs optimize process control and enhance operational efficiency. These sensors provide real-time industrial process monitoring and control, which boosts output and decreases downtime.

In industrial processes, OFSs can keep an eye on variables like temperature, pressure, and chemical composition. AI algorithms analyze the sensor data to optimize process conditions and ensure product quality [22]. By tracking the condition of the equipment and anticipating faults before they happen, these sensors make predictive maintenance possible. Artificial intelligence (AI) systems examine both past and current data to find trends that point to upcoming failures, enabling prompt maintenance and minimizing unscheduled downtime [9].

d. **Smart Agriculture**

The integration of AI and OFSs offers promising opportunities for advancing various fields. Another significant area is smart agriculture. AI-enabled OFSs can be utilized for real-time soil health monitoring, assessing parameters such as moisture and nutrient levels to optimize irrigation and fertilization strategies. Additionally, these sensors can facilitate crop health diagnostics by employing hyperspectral imaging to detect early signs of plant diseases through analysis of chlorophyll content and other biochemical markers.

e. **Smart Cities**

In the context of smart cities, AI-enhanced OFSs can play a crucial role in structural health monitoring. These sensors can dynamically monitor the load on bridges and buildings, using AI to predict structural responses to environmental changes like temperature fluctuations and seismic activities. Another application is in smart pavements, where OFSs can track stress and wear, allowing for optimized maintenance schedules and enhanced public safety.

f. **Telecommunications**

Telecommunications can also benefit from the integration of AI and OFSs. Real-time network monitoring can be enhanced by these sensors, which analyze data transmission quality, detect anomalies, and optimize bandwidth allocation dynamically. Additionally, exploring their potential in quantum communication could enhance cybersecurity measures in telecommunications through secure quantum key distribution.

g. **Energy Management Systems**

AI-enabled OFSs can monitor renewable energy sources, including wind turbines and solar panels, and provide real-time data to energy management systems so that energy production may be optimized based on environmental conditions. These sensors can also be crucial for smart grids, facilitating real-time data acquisition and analysis to improve load balancing and energy distribution.

h. **Security and Surveillance**

Security and surveillance applications can be significantly improved with AI-integrated OFSs. For instance, they can serve as intrusion detection systems where disturbances in the optical fibers trigger alerts, allowing for real-time surveillance analysis. Furthermore, these sensors can monitor environmental changes in sensitive areas, predicting and responding to potential disasters like landslides or floods.

i. **Autonomous Vehicles**

In the realm of autonomous vehicles, AI-enhanced OFSs can assist with navigation and sensing by providing real-time environmental data to enhance obstacle detection and navigation capabilities. They can also monitor critical vehicle components, with AI predicting maintenance needs based on wear patterns and performance data.

j. **Biomedical Research**

Biomedical research stands to gain from this integration as well. For example, lab-on-a-chip applications can be developed using AI and OFSs to create portable diagnostic tools for rapid disease detection in remote settings. Additionally, neurotechnology applications can leverage these sensors in brain–computer interfaces, improving communication and control for individuals with disabilities by analyzing neural signals. Rachana et al. discussed recent advances in fiber sensors based on photonic crystal for biomedical applications [19]. A survey of the several photonic crystal fiber (PCF) shapes employed in biomedical sensors is presented in this work. The paper's main goal is to demonstrate the various uses of biomedical sensors, including the detection of DNA, proteins, blood, and glucose. Using a variety of backdrop materials, the study examines the sensitivity and confinement loss of these variably shaped PCFs.

k. **Space Exploration**

Lastly, space exploration can benefit from AI-integrated OFSs for monitoring environmental conditions within spacecraft, ensuring optimal living conditions for astronauts. These sensors could also be used to analyze soil and atmospheric conditions on other planets, employing AI to process complex datasets for remote exploration.

AI-based OFSs offer significant advancements in various applications, from healthcare diagnostics to environmental monitoring, structural health monitoring, and industrial automation. By integrating AI techniques, these sensors provide enhanced data processing, intelligent decision-making capabilities, and real-time analysis, paving the way for smarter and more efficient sensing systems (Sasy [23]).

5 Case Studies

a. **Structural Health Monitoring with AI-Enhanced OFSs**

In large infrastructure projects like bridges and skyscrapers, monitoring structural integrity is critical for safety and longevity. Traditional OFSs were used to monitor strain, temperature, and vibrations, but their performance was often hindered by environmental noise and complex data patterns. By integrating ML techniques, especially SVM and neural networks, the accuracy of signal interpretation was greatly improved.

In this application, ML algorithms were trained to recognize patterns in the sensor data, distinguishing between regular structural vibrations and those indicating potential damage. For example, an AI-enhanced OFS system deployed on a bridge was able to predict cracks in its structure with high accuracy by continuously learning from historical data. This predictive capability helped maintenance teams perform repairs before critical damage occurred, extending the structure's life and reducing downtime. A study by Wang et al. investigated real-time respiratory monitoring and sitting posture recognition device based on a wearable double fishtail-shaped FBG wearable device [33].

In healthcare, **Wearable Devices** like Fitbit and Apple Watch integrate AI to enhance sensor accuracy. These devices monitor vital signs and use AI to filter noise and detect anomalies in heart rate patterns. This has resulted in quicker identification of arrhythmias, improving patient outcomes significantly.

b. **Pipeline Monitoring with Distributed Acoustic Sensing (DAS)**

DAS systems are widely applied for monitoring pipelines for potential leaks, tampering, and mechanical failures. However, differentiating between normal pipeline sounds and abnormal events such as a leak posed a challenge. By integrating DL models like CNNs and RNNs, the performance of these sensors was drastically enhanced.

In a real-world scenario, an AI-enhanced DAS system was deployed along a major oil pipeline. The CNN models were used to process acoustic signals and extract meaningful features, while the RNNs analyzed time-series data to track sound patterns over time. This combination enabled the system to detect leaks with a much higher degree of accuracy and reduced the occurrence of false positives. AI's ability to differentiate between operational noise and leak-related sounds led to early detection and immediate corrective actions, saving costs associated with leaks and environmental damage.

c. **Medical Diagnostics using AI-Driven Optical Coherence Tomography (OCT)**

OCT, an imaging method used to capture high-resolution cross-sectional images of tissues, was integrated with deep learning algorithms to improve diagnostic accuracy. AI-enhanced OCT systems have shown tremendous success in diagnosing retinal conditions such as macular degeneration and diabetic retinopathy.

In this real-world application, a deep learning model trained on large datasets of OCT images was used to analyze new scans. The AI system could detect subtle changes in the retinal structure that were often missed by human examiners. By integrating AI, doctors were able to diagnose diseases earlier, plan treatments more effectively, and significantly improve patient outcomes. The system also reduced the time required for diagnosis and improved accuracy, thus minimizing invasive diagnostic procedures.

d. **Environmental Monitoring for Air Quality**

In environmental monitoring, optical fiber sensors integrated with machine learning algorithms were used to measure air quality in urban areas. These sensors detect changes in light properties caused by particulate matter and gases in the air. However, real-time interpretation of the data in varying environmental conditions was challenging due to the noisy and dynamic nature of the data. AquaMonitor, a company using AI to analyze sensor data from water quality sensors, can detect contamination events in real time. By integrating machine learning, they improved the detection accuracy of harmful substances, enabling rapid response and minimizing public health risks.

Incorporating unsupervised learning methods such as PCA and k-means clustering will enable the sensors to automatically classify and sort various contaminants in real time. These models helped analyze large streams of data, identifying harmful air quality trends faster than traditional methods. In a real-world application in a major city, this AI-enhanced system helped local authorities respond quickly to pollution spikes, improving public health by issuing timely warnings and enforcing environmental regulations.

e. **Smart Agriculture**

AI-powered sensors in precision agriculture have improved crop monitoring. For instance, **CropX** uses soil sensors that analyze moisture levels. The AI algorithms process this data to optimize irrigation schedules, leading to a 20% reduction in water usage and a 10% increase in crop yields, demonstrating enhanced sensor performance through intelligent data utilization.

f. **Industrial IoT**

Companies like **Siemens** implement AI in industrial sensors for predictive maintenance. To anticipate errors before they happen, their systems evaluate data from sensors installed in machinery. This approach has reduced downtime by up to 30%, showcasing the direct enhancement of sensor functionality through AI integration.

6 Challenges and Potential Solutions

The integration of AI with OFSs holds substantial promise for advancing sensor technology and expanding its applications. Despite these promising developments, several challenges remain which need to be addressed. The integration of AI requires large, diverse datasets, which are complex to develop and maintain, posing challenges for ensuring accurate AI model performance. Furthermore, implementing AI models necessitates substantial computational resources, leading to high costs and significant energy consumption, with a need for efficient power usage at the edge. Technical integration issues, such as aligning AI algorithms with existing sensor systems and communication interfaces, complicate real-time data processing, particularly on a large scale. The lack of standardization in both sensor technologies and AI methodologies hampers widespread adoption, emphasizing the need for established norms and compatibility standards. Finally, ethical considerations and regulatory compliance concerning data privacy and security are critical to ensure that AI applications in sensing technologies are implemented responsibly and without compromising user trust.

A complex strategy combining cutting-edge technologies and creative tactics is needed to address these issues. Improving data accuracy can be achieved through the integration of ML and DL techniques. These AI algorithms can filter noise, enhance signal processing, and refine data interpretation in real time. For example, deep learning models like CNNs can be used to automatically extract meaningful features from noisy sensor data, while RNNs can track time-series data to predict changes in monitored environments. Additionally, using data augmentation techniques can help address the issue of limited training data by synthetically generating more samples, thus improving the robustness and reliability of AI models.

The cost challenges of OFS systems, especially in large-scale deployments, can be mitigated through advances in manufacturing technologies and materials. The development of cost-effective polymer-based optical fibers could reduce production expenses while maintaining performance. Moreover, the rise of edge computing and AI-driven analytics reduces the need for centralized, high-cost processing infrastructure by enabling data processing at the sensor level. This decentralization can further lower operational costs while ensuring real-time data analysis and reducing latency.

To address the challenge of environmental durability, particularly in harsh conditions, the use of protective coatings and encapsulation techniques can safeguard sensors from damage due to temperature fluctuations, moisture, and mechanical stress. Researchers are also exploring self-healing materials that can repair minor damages to optical fibers automatically, enhancing their longevity in rugged environments. Moreover, AI algorithms can improve durability by enabling predictive maintenance through continuous monitoring, reducing unexpected failures and extending sensor lifespans. By integrating AI, advanced materials, and manufacturing innovations, the challenges of data accuracy, cost, and environmental durability in optical fiber sensors can be effectively addressed.

7 Emerging Trends in AI for Enhanced Optical Fiber Sensors

Recent advancements in artificial intelligence (AI) are poised to significantly impact the development of optical fiber sensors (OFSs), particularly in areas such as data processing, real-time monitoring, and predictive maintenance. As AI technologies evolve, their integration with OFS is opening new possibilities for improving sensor accuracy, efficiency, and robustness across various applications.

One emerging trend is the use of DL models, particularly CNNs and RNNs. These models offer powerful tools for handling the complex, high-dimensional data generated by OFSs. CNNs, originally developed for image recognition tasks, are now being applied to optical sensor data to enhance feature extraction and pattern recognition. For instance, CNNs can identify subtle changes in light intensity or phase in OFS data, which can be critical for applications like structural health monitoring or environmental sensing. In dynamic contexts like pipeline monitoring, where continuous data streams must be processed and evaluated quickly, RNNs—especially LSTM networks—are well suited for processing time-series data. This allows for more accurate real-time monitoring.

Another notable advancement is in the field of edge AI and distributed intelligence. As the volume of data generated by OFS systems increases, there is a growing need to process this data locally rather than relying on centralized computing systems. By processing data closer to the source, edge AI speeds up decision-making and lowers latency. This is particularly significant for applications requiring real-time response, such as in industrial automation or autonomous vehicle navigation. By deploying AI models on edge devices, OFSs can perform advanced signal processing, anomaly detection, and even predictive analytics without requiring constant data transmission to a remote server. This trend lowers the energy consumption related to cloud-based data processing while also improving response times.

The integration of unsupervised learning techniques is also gaining traction in OFS development. Unlike supervised models that rely on labeled datasets, unsupervised learning algorithms can detect patterns in unlabeled data, making them ideal for exploring new environments or applications where predefined labels are not available. Techniques such as k-means clustering and PCA are being used to identify anomalies and group similar data points in large-scale OFS deployments. For example, in environmental monitoring, these models can automatically classify different types of pollutants or changes in water quality based solely on the data collected, without requiring predefined labels. This flexibility allows for more adaptive and autonomous sensing systems.

Advancements in reinforcement learning (RL) are further expanding the potential of AI-enhanced OFSs. RL models are particularly useful for dynamic, adaptive systems that need to continuously improve based on real-time feedback. In the context of OFSs, RL can be used to optimize sensor networks by learning from environmental data and adjusting sensing parameters accordingly. For instance, in smart cities or industrial IoT applications, RL can help optimize sensor placement, energy usage, and data transmission protocols, maximizing efficiency while maintaining high performance.

AI's role in predictive maintenance is another key advancement affecting OFS development. ML models may anticipate when a sensor or system is likely to malfunction by evaluating prior sensor data. This enables proactive maintenance and minimizes downtime. In sectors like oil and gas, where OFSs are used to check pipelines for leaks or structural damage, this predictive capability is especially useful.

AI-driven predictive maintenance reduces operational costs and enhances safety by ensuring that potential issues are addressed before they become critical.

Moreover, the development of quantum computing is expected to further enhance AI's impact on OFSs. Quantum computing offers unparalleled processing capabilities, which can significantly accelerate the training of complex AI models. As quantum computers become more accessible, they could be used to develop more sophisticated AI algorithms for OFSs, enabling faster and more accurate analysis of large datasets, particularly in fields like environmental monitoring, medical diagnostics, and aerospace.

Lastly, the integration of AI with IoT platforms is creating a more interconnected sensing environment. By linking OFS with other sensor systems through IoT networks, AI algorithms can analyze data from multiple sources simultaneously, providing more comprehensive insights. This holistic approach is particularly beneficial in applications like smart infrastructure, where multiple types of sensors are deployed to monitor various aspects of a system, like temperature, strain, and vibration. AI can process this multisensory data to provide more accurate and actionable information.

In summary, emerging trends and advancements in AI, including deep learning, edge AI, unsupervised learning, reinforcement learning, quantum computing, and IoT integration, are driving the next generation of optical fiber sensors. These innovations are improving sensor performance in terms of data accuracy, real-time processing, and predictive maintenance, making AI-enhanced OFS a key technology for future smart systems.

8 Conclusion

The integration of AI with OFSs represents a transformative advancement in sensing technology, enhancing sensitivity, detection accuracy, and data processing capabilities. This chapter has detailed how AI algorithms improve sensor performance, starting from the fundamental principles of optical fiber sensors, which rely on light modulation to measure external variables. Various sensor subclasses, categorized by configuration and sensing elements, elucidate their operational mechanisms.

Optical fiber sensors find applications across multiple domains, including structural health monitoring, environmental assessment, biomedical fields, industrial processes, aerospace, and defense, showcasing their versatility and precision. The chapter also addressed key challenges such as data accuracy, system complexity, cost, environmental durability, and compatibility, which must be overcome to optimize global use and performance.

Significant emphasis was placed on AI's role in signal processing, highlighting how ML, DL, and NNs enhance signal quality, pattern recognition, and real-time decision-making, thus improving overall utility and precision. The future of AI-assisted optical fiber sensors looks promising, with ongoing advancements expected to drive the development of intelligent sensing systems. However, addressing barriers such as security, ethical, and data privacy considerations is crucial to fully realize the potential of this interdisciplinary approach.

References

1. N.H. Al-Ashwal, K.A.M. Al Soufy , M.E.E. Hamza , N. Cennamo, M.J. Deen, J. Lee, S. Mukhopadhyay, S. Morais, N.H. Al-Ashwal, K.A.M. Al Soufy, M.E. Hamza, M.A. Swillam, Deep learning for optical sensor applications: a review. Sensors **23**(14), 6486 (2023). https://doi.org/10.3390/S23146486
2. Applications, P. T.-I. T. on I. & 1991, undefined. (n.d.). Intrinsic fiber-optic sensors. *Ieeexplore.Ieee.Org*. Retrieved July 8, 2024, from https://ieeexplore.ieee.org/abstract/document/67537/?casa_token=R_rAYSI6fxAAAAAA:K_cNM4pZaLf_H88NgUpQXr1Hhn-3hvkIS6_xvzF5qyuY-s7_jitS9EPh4Hulaqd_FlQP9scHP8E
3. A. Bogomolov, A. Evseeva, E. Ignatiev, V. Korneev, New approaches to data processing and analysis in optical sensing. TrAC Trends Anal. Chem. **160**, 116950 (2023). https://doi.org/10.1016/J.TRAC.2023.116950
4. J. Castrellon-Uribe, (2012). *Optical fiber sensors: an overview*. https://openresearchlibrary.org/ext/api/media/1daf3b26-0783-4ca1-b391-c5b025ba98a9/assets/external_content.pdf
5. A. Datta, V. Raj, V. Sankar, S. Kalyani, express, B. S.-O. & 2021, undefined. (n.d.). Measurement accuracy enhancement with multi-event detection using the deep learning approach in Raman distributed temperature sensors. Opg. Optica. Org. Retrieved July 8, 2024, from https://opg.optica.org/abstract.cfm?uri=oe-29-17-26745
6. H.S. Efendioglu, T. Yildirim , O. Toker, Advanced image processing and artificial intelligence based approaches to fiber optic statistical mode sensor design **7982**, 256–262 (2011). https://doi.org/10.1117/12.880054
7. M.M.A. Eid, Optical fiber sensors: review of technology and applications. Indonesian J. Electr. Eng. Comput. Sci. **25**(2), 1038–1046 (2022). https://doi.org/10.11591/IJEECS.V25.I2.PP1038-1046
8. K.T.V. Grattan, New developments in sensor technology—fibre and electro-optics. Measurement and Control **22**(6), 165–175 (1989). https://doi.org/10.1177/002029408902200602/ASSET/002029408902200602.FP.PNG_V03
9. B.E. Jones, Optical fibre sensors and systems for industry. J. Phys. E: Sci. Instrum. **18**(9), 770 (1985). https://doi.org/10.1088/0022-3735/18/9/007
10. S.B. Junaid, A.A. Imam, M. Abdulkarim, Y.A. Surakat, A.O. Balogun, G. Kumar, A.N. Shuaibu, A. Garba, Y. Sahalu, A. Mohammed, T.Y. Mohammed, B.A. Abdulkadir, A.A. Abba, N.A. Iliyasu Kakumi, A.S. Hashim, Recent advances in artificial intelligence and wearable sensors in healthcare delivery. Appl. Sci. **12**(20), 10271 (2022). https://doi.org/10.3390/APP122010271
11. E. Kavuncuoğlu, F. Yıldız, A.Ö. Optik, undefined. (n.d.). Artificial Intelligence (AI) algorithms for evaluation of optical fiber scintillation detector performance. Elsevier (2022). Retrieved July 8, 2024, from https://www.sciencedirect.com/science/article/pii/S0030402622001930
12. A. Leal-Junior, J. Silva, L. Macedo, A. Marchesi, S. Morau, J. Valentino, F. Valentim, M. Costa, The role of optical fiber sensors in the new generation of healthcare devices: a review. Sensors & Diagnostics (2024). https://doi.org/10.1039/D4SD00032C
13. B. Li, Artificial intelligence for optical fiber sensors (2023). https://dr.ntu.edu.sg/handle/10356/164277
14. C. Lyu, Z. Huo, X. Cheng, … J. J.-J. of L. & 2020, undefined. (n.d.). Distributed optical fiber sensing intrusion pattern recognition based on GAF and CNN. *Ieeexplore.Ieee.Org*. Retrieved July 8, 2024, from https://ieeexplore.ieee.org/abstract/document/9063486/?casa_token=PY9EeHqvQdQAAAAA:1xp-TSHSPzqS0n4Gdv3Bk_sXRPswHUWUvAoJd2EVi1Zm2SqJiZ4DnMbTmX2hDV5xI_0EkVHl5xE

15. J. Mata, I. de Miguel, R.J. Durán, N. Merayo, S.K. Singh, A. Jukan, M. Chamania, Artificial intelligence (AI) methods in optical networks: a comprehensive survey. Opt. Switch. Netw. **28**, 43–57 (2018). https://doi.org/10.1016/J.OSN.2017.12.006
16. I. Noiseux, W. Long, A. Cournoyer, M. Vernon, Simple fiber-optic-based sensors for process monitoring: an application in wine quality control monitoring. Appl. Spectrosc. **58**(8), 1010–1019 (2004). https://doi.org/10.1366/0003702041655476
17. X. Pan, S. Zhang, C. Zhang, N. Song, High-performance fully differential photodiode amplifier for miniature fiber-optic gyroscopes. Optics Express **27**(3), 2125–2141 (2019). https://doi.org/10.1364/OE.27.002125
18. S. Patel, H. Park, P. Bonato, L. Chan, M. Rodgers, A review of wearable sensors and systems with application in rehabilitation. J. Neuroeng. Rehabil. **9**(1), 1–17 (2012). https://doi.org/10.1186/1743-0003-9-21
19. M. Rachana, I. Charles, S. Swarnakar, S.V. Krishna, S. Kumar, Recent advances in photonic crystal fiber-based sensors for biomedical applications. Opt. Fiber Technol. **74**, 103085 (2022). https://doi.org/10.1016/J.YOFTE.2022.103085
20. S.C. Sajan, A. Singh, P.K. Sharma, S. Kumar, Silicon photonics biosensors for cancer cells detection—a review. IEEE Sens. J. **23**(4), 3366–3377 (2023). https://doi.org/10.1109/JSEN.2023.3235920
21. S.A. Saleah, S. Kim, J.A. Luna, R.E. Wijesinghe, D. Seong, S. Han, J. Kim, M. Jeon, Optical coherence tomography as a non-invasive tool for plant material characterization in agriculture: a review. Sensors **24**(1), 219 (2023). https://doi.org/10.3390/S24010219
22. J.L. Santos, Optical Sensors for Industry 4.0. IEEE J. Sel. Topics Quantum Electron. **27**(6) (2021). https://doi.org/10.1109/JSTQE.2021.3078126
23. Y.W. Sasy Chan, H.P. Wang, P. Xiang, Optical fiber sensors for monitoring railway infrastructures: a review towards smart concept. Symmetry **13**(12), 2251 (2021). https://doi.org/10.3390/SYM13122251
24. K. Sharma, M. Sharma, Optical biosensors for environmental monitoring: recent advances and future perspectives in bacterial detection. Environ. Res. **236**, 116826 (2023). https://doi.org/10.1016/J.ENVRES.2023.116826
25. D. Singh, M. Dahiya, R. Kumar, C. Nanda, Sensors and systems for air quality assessment monitoring and management: a review. J. Environ. Manage. **289**, 112510 (2021). https://doi.org/10.1016/J.JENVMAN.2021.112510
26. Q. Song, G. Yan, G. Tang, F.A.-M.S. Signal, and & 2021, undefined. (n.d.-a). Robust principal component analysis and support vector machine for detection of microcracks with distributed optical fiber sensors. *Elsevier*. Retrieved July 8, 2024, from https://www.sciencedirect.com/science/article/pii/S0888327020304052
27. Q. Song, G. Yan, G. Tang, F.A.-M.S. Signal, and & 2021, undefined. (n.d.-b). Robust principal component analysis and support vector machine for detection of microcracks with distributed optical fiber sensors. *Elsevier*. Retrieved July 8, 2024, from https://www.sciencedirect.com/science/article/pii/S0888327020304052
28. Technology, Y. R.-O. F. & 2006, undefined. (n.d.). Recent progress in fiber-optic extrinsic Fabry–Perot interferometric sensors. *Elsevier*. Retrieved July 8, 2024, from https://www.sciencedirect.com/science/article/pii/S1068520006000174
29. E. Udd, An overview of fiber-optic sensors. Rev. Sci. Instrum. **66**(8), 4015–4030 (1995). https://doi.org/10.1063/1.1145411
30. E. Udd, W.B. Spillman, (n.d.). Fiber optic sensors: an introduction for engineers and scientists. 597.
31. Using global existing fiber networks for environmental sensing. (n.d.). *Ieeexplore.Ieee.Org*. Retrieved July 6, 2024, from https://ieeexplore.ieee.org/abstract/document/9881563/?casa_token=Ox-ZOJLH3hsAAAAA:KymbrFUlX6JwRyRUUEe1qS_mIAiJT16ZuoTIh6xhvSzGkEl_BPoTCXGhDTA1Qiau-yBdBk4X
32. A. Venketeswaran, N. Lalam, J. Wuenschell, P.R. Ohodnicki, M. Badar, K.P. Chen, P. Lu, Y. Duan, B. Chorpening, M. Buric, Recent advances in machine learning for fiber optic sensor applications. Adv. Intell. Syst. **4**(1), 2100067 (2022). https://doi.org/10.1002/AISY.202100067

33. H. Wang, J. Zheng, Q. Nie, W. Li, Z. Wang, K. Xiao, X. Hu, S. Kumar, R. Min, Double-fishtail-shaped FBG wearable device for sitting posture recognition and real-time respiratory monitoring. IEEE Sens. J. **24**(7), 10164–10171 (2024). https://doi.org/10.1109/JSEN.2024.3367622
34. W. Zeng, H.G.-S. Transducers, & 2014, undefined. (2014). Optic fiber sensing IOT technology and application research. *Citeseer*, *180*, 16–21. https://citeseerx.ist.psu.edu/document?repid=rep1&type=pdf&doi=ccf9da51e3d7f2d04ac60d468a4c13ff40af8373
35. B. Zha, Z. Wang, L. Ma, J. Chen, H. Wang, X. Li, S. Kumar, R. Min, Intelligent wearable photonic sensing system for remote healthcare monitoring using stretchable elastomer optical fiber. IEEE Internet Things J. **11**(10), 17317–17329 (2024). https://doi.org/10.1109/JIOT.2024.3356574
36. J. Zhang, Y. Xiang, C. Wang, Y. Chen, S.C. Tjin, L. Wei, Recent advances in optical fiber enabled radiation sensors. Sensors **22**(3), 1126 (2022). https://doi.org/10.3390/S22031126
37. Y. Zhou, Y. Zhang, nan, Q. Yu, L. Ren, Q. Liu, Y. Zhao, Application of machine learning in optical fiber sensors. Measurement**228**, 114391 (2024). https://doi.org/10.1016/J.MEASUREMENT.2024.114391

AI Techniques for Signal Processing in Optical Fiber Sensors

Jyoti Katyal

Abstract AI techniques, from machine learning (ML) to the most recent developments in deep learning (DL), are widely utilized in optical communication and networks. This chapter explores advanced ML and DL methods and their applications in processing fiber optic sensors. It emphasizes the potential of these algorithms for the design and development of intelligent sensors. The chapter addresses the challenges and limitations of fiber optic sensors and how AI has addressed these issues. AI has significantly enhanced signal processing in optical fiber sensors by improving data analysis, increasing accuracy, and accelerating decision-making. AI techniques facilitate the effective management and interpretation of complex sensor data, resulting in more accurate and timely insights. Recent advancements in ML-based fiber sensors are reviewed, highlighting the problems encountered with conventional methods. Machine learning (ML) techniques have been applied to optimize various aspects of fiber optic sensors. In the past 20 years, the integration of machine learning (ML) and artificial intelligence (AI) has transformed various scientific fields, facilitated by tools like MATLAB, Scikit-Learn, and Tensor Flow. These advancements include Scientific Machine Learning (SciML) and physics-informed ML algorithms, which combine domain knowledge with data-driven models. Deep learning has notably improved pattern recognition, often matching human-level performance. This chapter highlights recent developments in ML and AI that enhance fiber optic sensing (FOS) applications and provide a thorough overview of ML techniques applied to optical fiber sensors focusing on various AI methods that enhance sensor performance. Additionally, it discusses future developments in this evolving research field.

Keywords Fiber optic sensor · Artificial intelligence · Machine learning/deep learning · Signal processing

J. Katyal (✉)
Amity Institute of Applied Science, Amity University, Noida, India
e-mail: jyoti_106@yahoo.co.in

V. Bhardwaj et al. (eds.), *Optical Fiber Sensors and AI*, Progress in Optical Science and Photonics 34, https://doi.org/10.1007/978-981-96-6158-9_4

1 Introduction

Artificial intelligence (AI) entities and systems can perform tasks similar to learning and decision-making by mimicking biological processes, particularly understanding human functions. AI applications, including healthcare, industrial equipment, environmental, networking, are nearly everywhere. These AI-based technologies are transforming our daily lives, enhancing productivity, safety, and health, and even influencing how we entertain ourselves and communicate [1–3]. For the most part, AI does not provide fully autonomous systems. Instead, it enhances existing applications, databases, and environments by adding knowledge and reasoning, making them more user-friendly, intelligent, and responsive to environmental changes. Advancements in AI research enable us to tackle novel and larger scale problems, fostering research and innovation in almost every scientific discipline. For instance, the application of AI techniques to enhance the speed of telecommunication networks has been in focus over the past decades [4–6]. Optical communication systems have also adopted integrating AI methods into various aspects of photonic devices. Machine learning (ML), a branch of artificial intelligence (AI), has been pivotal in advancing the capabilities of optical fiber sensors. Nature photonics further explored this area with a special issue titled "Machine Learning of Light" [7]. Optics, a fundamental area in microphysics, exhibits complexities akin to the "black-box" nature of ML processes. Applying optics and photonics to machine learning tasks offers significant benefits, such as enabling rapid computations at very high frame rates while consuming minimal energy. The two main factors driving the current application of ML in photonics are discussed now [8]. One approach involves using AI techniques in hardware to design optical structures and devices with specific performance characteristics. Another approach involves using optical systems as either fully optical or hybrid statistical inference models for AI computations. Feldmann et al. [9] presented a scalable circuit architecture for photonic neural networks (NNs) by means of wavelength division multiplexing technology, resulting in a fully optical neural synaptic system. They use phase-change materials that encase neurons and use optical pulse signals to control the materials' transmittance and reflectivity through the use of controlled lasers. This system exhibits efficient pattern recognition in a fully optical framework, supporting both supervised and unsupervised learning. By utilizing the intrinsic speed and bandwidth of optical networks, this photonic NN makes it possible to process optical communication and visual data directly. But problems still exist since logic gates, calculations, and NNs are linear and cannot deal with nonlinear problems well. The implementation of an all-optical method is complicated by the unavailability of effective nonlinear optical processes as activation functions for several nodes. The integration of polymer optical fiber (POF) speckle sensors with smartphones to enhance human–computer interaction (HCI) is also one of the example [10]. This approach leverages the advantages of POF technology for gesture recognition and biometric identity verification. POF speckle sensors work by analyzing the interference patterns (speckle patterns) created by the interaction of light with a surface. These patterns are sensitive to minute changes, which can be used to detect gestures and biometric traits. Other article which presents a novel insole sensing system designed for plantar healthcare applications integrates artificial intelligence (AI) with advanced sensing technology to monitor and analyze various aspects of foot health [11–13]. The system aims to provide comprehensive, real-time feedback for both medical and preventative care. The insole system incorporates AI algorithms to process data from various sensors embedded in the insole. This AI-driven approach enhances the system's ability to interpret complex data related to foot biomechanics and health. The development and application of portable optical fiber biosensors integrated with smartphones has advanced the technology. The article focuses on how optical fiber biosensors can be integrated with smartphones, leveraging the computational power and connectivity of mobile devices to enhance biosensing capabilities [14, 15]. It explores the technological advancements, practical applications, and the challenges associated with this integration, aiming to provide a comprehensive understanding of the field. Also a novel wearable device featuring a double-fishtail-shaped fiber Bragg grating (FBG) sensor designed for monitoring sitting posture and respiratory patterns in real time aims to enhance health monitoring and ergonomics by providing accurate and continuous data. The use of silicon photonics biosensors in the detection of cancer cells explores the advancements in silicon photonics technology and its application in cancer diagnostics [16, 17].

Total internal reflection-based optical fiber sensors have many benefits, such as high sensitivity, low loss, affordability, user-friendliness, immunity to electromagnetic interference, and appropriateness for online and remote monitoring [18, 19]. Based on how they are configured, these sensors are divided into pointwise and distributed types. By adjusting the surface material, discrete sensors may perform multi-parameter sensing and get high measurement accuracy. They are widely researched for applications in wearable healthcare and industrial process monitoring, facilitating plug-and-play online sensing and monitoring [18, 19]. Distributed sensors are gaining popularity due to their ability to continuously monitor objects over long optical fiber distances [19]. Traditional demodulation techniques face challenges such as nonlinear interference and limited measurement range with optical fiber sensors. ML addresses these issues by mitigating cross-sensitivity in some sensors and enhancing overall performance. In industrial settings, ML aids in optimizing optical device development by enabling transferable models across similar production devices, thereby streamlining large-scale manufacturing and assembly of optical products [20]. Significant advancements have been achieved in applying machine learning and optical sensors across different fields [21, 22]. For example, an optical immunosensor has been developed that combines machine learning with signal transduction, utilizing microspheres for encoding [23]. This innovation enables rapid and simultaneous detection of antibiotics in milk through polystyrene nanoparticles. These nanoparticles, which can be easily modified in size and quantity, carry multiple signal probes, facilitating straightforward detection of various antibiotics in a single-step signal conversion process. Also, a series of label-free colorimetric sensors based on Fabry–Pérot films featuring an organic–metallic structure to detect volatile organic compounds and monitor food spoilage have been used. ML analyzing sensor data, achieving a high classification accuracy of 92% at a concentration of 100 ppm for volatile organic compounds. This approach introduces a cost-effective platform of compact colorimetric sensors capable of assessing food quality, changes, and deterioration. Such technology holds promise for advancing smart labeling and packaging solutions [24].

Several studies in the literature have reviewed fiber optic sensing systems, as well as analyzed and discussed signal processing and machine learning methodologies for specific sensing technologies and application domains [25–30]. This work aims to offer a comprehensive review of AI-enabled fiber optic sensing (FOS) technologies by examining major trends in the field without focusing on a specific sensor or application. Initially, we address the challenges and limitations of FOS and explore the opportunities for developing machine learning solutions. Recent advancements in machine learning and AI could help overcome these challenges. We also highlight some of the latest AI techniques used to enhance FOS system processing. Finally, we summarize and propose future directions in this area.

2 Challenges and Limitations of FOS: Opportunities for ML and AI

Fiber optic sensing systems are advanced technologies that use optical fibers to measure various physical parameters. Pointwise FOS systems and distributed fiber sensors are major two sensing system. Pointwise FOS system has fiber Bragg grating sensor which has been extensively studied for monitoring various parameters, including temperature, strain, refractive index, pressure, curvature, and chemical composition. When there is a change in the surrounding temperature or strain, the grating period changes linearly, causing the reflected spectral wavelength to shift according to the applied conditions [31, 32]. Distributed fiber sensors provide innovative technology for spatially distributed measurements along the entire length of a fiber optic cable, utilizing Rayleigh, Brillouin, and Raman scattering. By analyzing the backscattered signal's frequency, intensity, and phase, it is possible to measure strain, temperature, or vibration over distances of tens of kilometers. Three scattering components occur within the optical fiber: Rayleigh, Brillouin, and Raman scattering [33, 34]. The expansion and adoption of fiber optic sensing (FOS) face various challenges and limitations, extensively explored and documented in the literature, Fig. 1. These challenges serve as a fertile ground for machine learning (ML) problems. By discussing these obstacles, we aim to inspire the growing AI scientific community to innovate ML solutions capable of overcoming these challenges effectively.

4.2.1 Data Availability: Unlike the extensive datasets used in computer vision and audio recognition (e.g., ImageNet for labeled images), most proposed ML frameworks for FOS applications lack publicly accessible datasets. However, acquiring additional distributed optical fiber sensor data is expensive and requires permission.

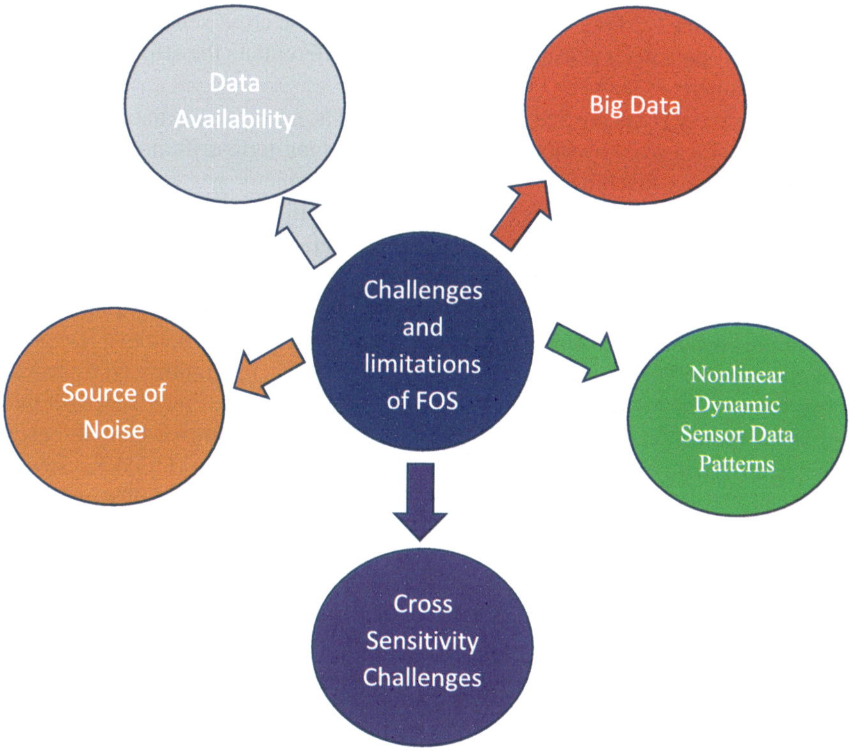

Fig. 1 Diagrammatic representation of challenges and limitations of FOS

4.2.2 Big Data: Distributed fiber optic sensing (DFOS) systems are known for generating substantial amounts of data due to their extensive sensing ranges and relatively small spatial resolutions [35]. The data output from Optical Frequency Domain Reflectometry (OFDR) systems scales prohibitively with the sensing range, limiting their practical range to a few hundred meters. Brillouin Optical Time-Domain Analysis (BOTDA) systems, while more accurate with smaller frequency steps, can also lead to big data challenges. Data compression algorithms are proposed to analyze this.

4.2.3 Sources of Noise: Low signal-to-noise ratio (SNR) is a common issue affecting many DFOS technologies. While earlier efforts aimed to resolve this problem through hardware improvements, recent work has shifted toward advanced machine learning-based data denoising algorithms [36, 37].

4.2.4 Cross-Sensitivity Challenges: A prevalent challenge with various point, quasi-distributed (such as FBGs), and distributed fiber optic sensing (DFOS) systems (including OFDR and BOTDA) is their susceptibility to cross sensitivity from multiple physical parameters. In absorption-based point sensors used for gas and chemical detection, cross sensitivity often arises between the target analyte and other ambient chemicals or environmental factors, such as temperature and humidity.

Advanced machine learning techniques can address these cross-sensitivity issues by enabling multi-parameter sensing, which helps to differentiate the effects of various physical parameters [38–41].

4.2.5 Nonlinear Dynamic Sensor Data Patterns: The time-varying nature of sensor responses, characterized by short-term transients or long-term drift, presents a challenge for applying machine learning (ML). Semiconductor gas sensors, whether electrical or optical, often exhibit rapid transients that differ significantly from their steady-state or long-term responses. Sensor drift refers to irreversible changes in sensor response over time or due to environmental exposure, typically requiring sensor design improvements to mitigate. However, ML can potentially exploit sensor data features that are insensitive to drift [42] and can reduce such data to a few principal component (PC) scores. Scores associated with PCs less influenced by wavelength bands prone to drift can exhibit less drift themselves. The effectiveness of this approach depends on how distinct the sensing response is from wavelength bands dominated by drift. In Optical Frequency Domain Reflectometry (OFDR) systems, changes in wavelength over time are nonlinear, which can degrade system spatial resolution. Robust nonlinear compensation techniques are essential, although current methods are most effective over short sensing ranges [43].

3 Recent Developments in ML with Relevance to FOS

Over the past 20 years, machine learning (ML) and artificial intelligence (AI) tools have become widely utilized across various scientific disciplines. This trend has been facilitated by software platforms, like MATLAB, Scikit-Learn, and Tensor Flow, which provide comprehensive ML modeling solutions. MATLAB known for its strong numerical computing capabilities is often used in sensor data processing for tasks such as signal filtering, data visualization, and implementing algorithms like Fourier transforms or Kalman filters. It is especially powerful in the areas of control systems, robotics, and embedded systems. For example, MATLAB can be used to process sensor data from autonomous vehicles to detect objects, measure distances, or even correct sensor noise. Scikit-Learn Python library is widely used in machine learning for tasks such as classification, regression, clustering, and anomaly detection. In the context of sensor data processing, Scikit-Learn is applied to build models that can detect patterns in large datasets generated by sensors. For instance, it might be used in predictive maintenance, where sensor data from industrial equipment is analyzed to predict potential failures. Tensor Flow is a deep learning framework that can handle large datasets and complex models, such as neural networks. In sensor data processing, Tensor Flow can be used to design models that recognize more abstract patterns, such as image or sound recognition. For example, Tensor Flow might be employed in the healthcare sector to analyze sensor data from wearable devices to monitor vital signs and detect anomalies like irregular heartbeats.

AI algorithms have significantly improved the performance of classical fiber optic sensors, addressing several key shortcomings. Fiber optic sensors are widely used in various industries due to their high sensitivity, immunity to electromagnetic interference, and ability to transmit data over long distances. However, they face challenges such as noise, limited resolution, and difficulty in processing large datasets. AI algorithms have provided solutions to these issues, enhancing the overall capabilities of these sensors.

Some of the key shortcomings addressed by AI in fiber optic sensors are given below:

Noise Reduction and Signal Enhancement: Fiber optic sensors are prone to various types of noise, such as thermal noise, shot noise, and phase noise, which can affect the accuracy and reliability of the sensed data. Classical filtering techniques often struggle to separate the useful signal from the noise effectively. AI-based techniques like machine learning (ML) and deep learning (DL) have been used to develop sophisticated noise reduction algorithms. For instance, convolutional neural networks (CNNs) can automatically learn to filter out noise while preserving relevant signal characteristics. Additionally, unsupervised learning models such as autoencoders have been employed for denoising sensor data without requiring large labeled datasets. In distributed acoustic sensing (DAS), AI algorithms have been used to filter out background noise and improve the detection of weak acoustic events. This has led to better performance in applications such as seismic monitoring, pipeline security, and intrusion detection. For example, a CNN-based approach significantly improved the detection of earthquakes and other seismic events using DAS systems by reducing the noise interference.

Improved Data Resolution: Classical fiber optic sensors, such as fiber Bragg grating (FBG) sensors, often have limitations in resolution due to the inherent properties of the sensor material and signal processing methods. Traditional approaches may not be able to resolve closely spaced events or accurately measure small changes in strain or temperature. AI techniques, such as reinforcement learning (RL) and hybrid AI systems, can enhance the resolution of fiber optic sensors by optimizing sensor interrogation methods. Reinforcement learning algorithms can dynamically adjust the sensor's operational parameters to achieve higher resolution. Additionally, AI can be used to develop new algorithms for accurately detecting and localizing strain or temperature changes from sensor data. A fiber Bragg grating sensor network equipped with AI-based processing algorithms was able to increase the spatial resolution of strain measurements in structural health monitoring applications. In this instance, machine learning algorithms were trained to interpret subtle changes in the optical signal, resulting in more precise strain localization and improved safety monitoring in civil engineering projects.

Handling Large Data Volumes: Distributed fiber optic sensing systems, such as distributed temperature sensing (DTS) and DAS, generate massive amounts of data, especially when used for real-time monitoring over long distances. Classical data processing techniques often struggle to handle these data volumes effectively, leading to bottlenecks in real-time analysis. AI algorithms, particularly deep learning models, excel at processing large datasets. With the use of parallel processing techniques and advanced neural networks, AI can handle and analyze vast amounts of data in real time. Furthermore, AI models can learn from historical data and provide predictive insights, reducing the need for constant human intervention. In oil and gas pipeline monitoring, DAS systems produce large volumes of acoustic data that must be analyzed to detect leaks or potential damage. AI algorithms have been implemented to process this data in real time, identifying anomalies that suggest pipeline issues. This not only accelerates the detection process but also reduces the likelihood of false positives, making the monitoring process more efficient.

Enhanced Pattern Recognition and Event Classification: Classical fiber optic sensors may struggle with recognizing complex patterns or classifying events in real time, particularly in environments where multiple events occur simultaneously. The traditional approach often requires predefined models that may not adapt well to changing conditions. AI, especially deep learning techniques such as recurrent neural networks (RNNs) and long short-term memory (LSTM) networks, can be applied to recognize and classify patterns in fiber optic sensor data. These models can learn temporal dependencies and detect complex patterns that are difficult to model using classical approaches. AI can also enable the classification of multiple events in parallel, improving the sensor's ability to operate in dynamic environments. In perimeter security applications, fiber optic sensors are used to detect and classify different types of intrusions, such as walking, digging, or vehicle movement. AI-based models have been trained to recognize these different intrusion types with high accuracy, even in challenging conditions where traditional models would struggle. For example, an AI-enhanced DAS system deployed along the perimeter of a critical infrastructure site successfully distinguished between human footsteps and vehicle movements, significantly improving the security of the site.

Furthermore, advancements in ML algorithms have produced new tools that seamlessly integrate domain-specific knowledge with data-driven ML models [44]. The rise of deep learning has notably improved pattern recognition models, often reaching performance levels akin to human capabilities. The following sections will provide a detailed overview of recent developments in ML and AI, which are expected to significantly advance the field of ML/AI-enhanced fiber optic sensing (FOS) applications.

4.3.1 Multiparametric Sensing: Multiparametric sensing involves the simultaneous measurement and differentiation of multiple physical parameters by a sensor. This is particularly challenging for fiber optic sensors (FOS) due to cross sensitivity, where different physical factors can interfere with each other's readings. Traditional sensors, like Fiber Bragg Grating (FBG) sensors, often face issues where temperature, strain, and pressure can overlap and affect measurements. To address these challenges, advanced machine learning (ML) techniques are employed. Random Forest models and deep learning architectures, such as stacked Autoencoders (AEs), help separate and analyze the effects of different parameters from sensor responses. This approach improves the accuracy and reliability of measurements in various applications, including structural health monitoring, environmental sensing, and industrial processes. Recent advancements include using sidelobe analysis in FBG spectra to distinguish between strain and temperature, and integrating deep learning with fiber-based Brillouin Optical Time-Domain Analysis (BOTDA) systems and Large-Effective-Area Fiber (LEAF) sensors. These innovations enhance the capability of fiber optic sensors to provide precise measurements in complex environments [45].

4.3.2 Accelerated Signal Processing: Acceleration of data processing is critical for handling the large volumes of data generated by Brillouin Optical Time-Domain Analysis (BOTDA) technologies. Various approaches have been proposed in the literature [46–49], categorized into 1) nonlinear curve-fitting/regression-based methods and 2) supervised machine learning (ML) modeling approaches. Regression-based methods offer advantages such as physics-based model selection and interpretability. However, these approaches can become cumbersome when processing datasets from long-range BOTDA systems (over 50–100 km sensing range). In contrast, supervised ML models significantly enhance processing speed. For instance, Zhang et al. [50, 51] introduced a radial basis function neural network (RBFNN) algorithm for BOTDR systems, demonstrating improved data processing speed and precise Brillouin Frequency Shift (BFS) measurements compared to traditional methods like Levenberg–Marquardt nonlinear least squares, cubic spline interpolation, and polynomial curve-fitting. ML-based approaches have also shown promise. Silva et al. [52] utilized ML for characteristic extraction, regression, and signal reconstruction in a Brillouin-based temperature sensing system, showcasing efficient data processing steps. Wu et al. [48] applied a support vector machine (SVM) for high-speed temperature measurement in BOTDA systems, achieving a 100-fold improvement in processing speed compared to conventional techniques. Wang et al. [53] implemented deep neural networks (DNN) for discriminating strain and temperature measurements using a Large-Effective-Area Fiber (LEAF) sensing fiber, significantly reducing data processing times compared to conventional fitting methods. Additionally, Azad et al. [54] proposed an artificial neural network (ANN)-based technique to mitigate Brillouin spectrum linewidth effects on measurement accuracy in a 41 km sensing fiber, reducing the required data points for accurate signal processing. These advancements underscore the effectiveness of ML and AI techniques in accelerating data processing for BOTDA technologies, offering improved speed, stability, reliability, and precision compared to traditional methods. Another artificial neural network (ANN)-based signal processing framework was developed using radial basis function (RBF) networks [55]. This framework employs a shallow neural architecture, requiring significantly fewer parameters compared to deep architectures. However, a major drawback of these neural network algorithms is the risk of overfitting and their deterministic approach to model inversion.

4 AI Techniques for Signal Processing

Artificial Intelligence (AI) has greatly advanced signal processing in optical fiber sensors by enhancing data analysis, improving accuracy, and speeding up decision-making processes. AI techniques help in effectively managing and interpreting complex sensor data, leading to more precise and timely insights.

The various methods used for fiber optic processing, Fig. 2, are as follows:

4.4.1 Random Forest: The Random Forest algorithm is a powerful machine learning method that builds multiple decision trees using random subsets of data and features. By aggregating the results of these trees, the algorithm reduces overfitting and improves prediction accuracy. For classification tasks, Random Forest uses majority voting, and for regression tasks, it averages the predictions of the trees. In the context

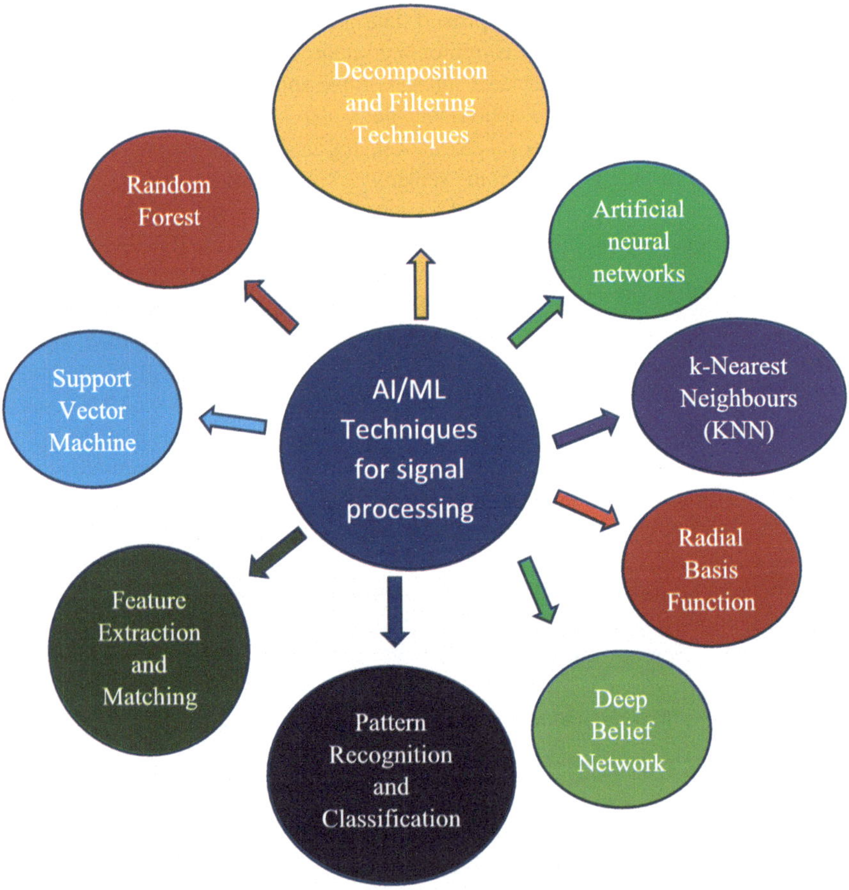

Fig. 2 Diagrammatical representation of AI/ML techniques for signal processing

of fiber optic sensors, Random Forest has been effectively applied to enhance performance. Zhaoyu Chen utilized this algorithm to develop a fatigue detection system for sports, while Jingxian Cui improved a two-dimensional vector displacement sensor using Random Forest, enhancing both its accuracy and range. These applications highlight Random Forest's potential for advancing point-based optical sensing, particularly in multi-parameter sensing scenarios [55, 56].

4.4.2 K-Nearest Neighbors (K-NN): The K-nearest neighbors (K-NN) algorithm is a straightforward, non-parametric supervised learning method used for classification and regression based on the proximity of data points. It classifies or predicts the value of a new data point by examining the classes or values of its K-nearest neighbors. In the context of Brillouin distributed fiber sensors, K-NN has been applied to extract Brillouin frequency shifts from the Brillouin gain spectrum. This method treats the extraction process as a model-free classification problem. The study reveals that K-NN outperforms traditional curve-fitting methods, with extraction accuracy influenced by factors such as signal-to-noise ratio (SNR) and frequency scanning step. Specifically, as the frequency scanning step increases or SNR decreases, performance declines. The K-NN algorithm can achieve a frequency shift uncertainty as low as 0.44 MHz with a 1 MHz frequency scanning step and an SNR of 11 dB [57].

4.4.3 Support Vector Machines: Support Vector Machine (SVM) is a versatile and robust machine learning algorithm used for classification, regression, and outlier detection tasks. Its effectiveness extends to various domains, including text and image classification, spam detection, handwriting recognition, and more. SVM excels in handling high-dimensional data and capturing complex, nonlinear relationships. In the context of optical fiber sensors for long-term structural health monitoring, a non-parametric approach is explored, leveraging SVM to analyze sensor data without assuming a fixed model. An advanced algorithm integrates Moving Principal Component Analysis (MPCA) with SVM. MPCA first identifies time-series eigenvectors from the sensor data, which are then processed by the SVM for classification or regression. This combination enhances performance by reducing dimensionality with MPCA and leveraging SVM's robust capabilities, leading to improved accuracy in detecting and assessing structural changes and potential damages over time.

4.4.4 Deep Belief Networks: Deep learning (DL) has been extensively applied in various optical sensor applications, enhancing accuracy and reducing noise levels. Optical sensors are increasingly recognized as a promising technology for modern intelligent sensing platforms, with application spanning process monitoring, quality prediction, pollution control, defense, security, and more. Despite their potential, optical sensors face significant challenges, including the management of large datasets, slow processing speeds, and high costs. Integrating DL systems with optical sensor technologies can help address these issues, improving data processing efficiency and reducing costs while leveraging the advanced capabilities of DL to enhance sensor performance. Key deep learning techniques and their uses include Convolutional Neural Networks (CNNs), Recurrent Neural Networks (RNNs), and Denoising Convolutional Neural Networks (DnCNNs). CNN is primarily used for processing visual data, CNNs can also be applied to sensor data where spatial patterns are important. They automatically learn complex features, making them effective for recognizing patterns and detecting anomalies in image-like data. RNNs are best suited for sequential data and are adept at capturing temporal relationships, which is valuable for time-series analysis, such as with fiber optic sensor data. They are useful for modeling sequences and forecasting future values based on historical data. DnCNNs are designed to reduce noise in data and enhance the quality and accuracy of sensor data by focusing on minimizing noise while maintaining crucial signal details.

4.4.5 Feature Extraction and Matching: It is another method used for processing of sensor having two main components: Principal Component Analysis (PCA) and Independent Component Analysis (ICA). PCA technique reduces dimensionality by identifying the most important features in the data, which enhances processing efficiency and lowers computational demands whereas ICA is used to separate mixed signals into their original components, making it effective for isolating distinct features from complex sensor data.

4.4.6 Decomposition and Filtering Techniques: Wavelet transform and adaptive filtering comes under this category. In fiber optic sensors, the wavelet transform is employed to examine signals across various scales or resolutions. This method is especially effective for processing non-stationary signals, which frequently occur in fiber optic systems due to their changing characteristics over time. By breaking down the signal into different frequency components, wavelet transforms facilitate a detailed analysis of transient features and anomalies that might be overlooked by other techniques. This enhances the accuracy and reliability of measurements in fiber optic sensor systems. Adaptive filtering is used to improve signal quality by adjusting to fluctuations in signal properties. This method employs filters that modify their settings in real time to effectively minimize noise and interference, which can change with different operating conditions and environments. By continuously adapting their filtering strategy, adaptive filters enhance the precision and clarity of the data gathered by fiber optic sensors, resulting in more dependable measurements and performance. Changsen Sun, Junxiu Lin performed an electronic adaptive filter using one-step residual error for the optical fiber pressure sensor [58].

4.4.7 Pattern Recognition and Classification: Various techniques are utilized for analyzing and interpreting sensor data, including Edge Detection, Nonlocal Means, and Anisotropic Diffusion. Edge Detection: Sheng Peng Wan's double-edge technique improves the measurement of Brillouin frequency shifts in fiber sensors by capitalizing on edge signal sensitivity to frequency changes. This approach demonstrates that double-edge techniques require lower signal-to-noise ratios (SNRs) to achieve high accuracy compared to conventional methods. Nonlocal Means: This method enhances feature extraction from noisy sensor signals by averaging similar data patches, effectively reducing noise. A combined denoising method integrating enhanced Discrete Wavelet Transform (DWT) with Nonlocal Mean (NLM) has been proposed to improve the quality of heartbeat signals from multimode fiber optic microvibration sensors. This method significantly increases output SNR across a wide range of input SNRs, outperforming traditional denoising techniques. Anisotropic Diffusion: This technique reduces noise while preserving essential edges and details in spatially distributed data. It is particularly useful for fiber optic sensors to maintain measurement accuracy and clarity, helping detect and analyze subtle changes and anomalies in sensor readings. These advancements enhance sensor data quality and diagnostic accuracy by improving noise reduction and feature preservation [59].

4.4.8 Radial Basis Function Neural Networks (RBFNNs) are a class of artificial neural networks that utilize radial basis functions for their activation functions. These networks are particularly effective for tasks like function approximation, pattern recognition, and time-series forecasting. Key features of RBFNNs include radial basis functions, layer structure, and training. Radial basis functions include the activation functions in RBFNNs which are usually Gaussian or other radial functions, dependent on the distance from a center point [60]. This capability allows RBFNNs to effectively model complex relationships in the data. Layer structure includes an RBFNN which comprises an input layer, a hidden layer where radial basis functions are applied, and an output layer. The hidden layer maps the input data into a higher dimensional space, facilitating pattern recognition and approximation. Training involves adjusting the parameters of the radial basis functions and the weights between the layers. This process often employs methods such as least squares or gradient descent.

4.4.9 Artificial neural networks: Recent advancements have seen the increasing use of Artificial Neural Networks (ANNs) to enhance the performance of optical fiber sensors for distributed strain and temperature measurements. ANNs, a key aspect of artificial intelligence, are composed of interconnected processing elements (neurons) with adjustable weights that generate outputs from given inputs. The model typically involves a linear combination of inputs followed by an activation function, such as the sigmoidal or hyperbolic tangent functions, with the latter often providing superior performance. Feedforward neural networks, also known as multilayer perceptron neural networks (MLPNNs), are the most commonly used type of ANN. In these networks, the first layer's neurons pass their outputs to the second layer's neurons without feedback. An ANN approach has been applied to predict the response of evanescent field absorption fiber sensors, demonstrating the network's ability to handle measurement errors and its significant role in developing intelligent sensors.

AI-enhanced signal processing has made significant advancements in fiber optic sensor technology which has greatly enhanced the practical relevance of the content. Here are a few detailed examples:

Smart Grid Technology: Fiber optic sensors, augmented by AI, are revolutionizing smart grid management. For instance, AI algorithms analyze data from fiber optic sensors deployed along power lines to detect and predict failures or outages. This real-time monitoring and predictive maintenance have led to increased reliability and efficiency in power distribution networks. An example is the deployment of AI-driven fiber optic sensors in smart grids in various cities, which has improved outage detection and response times.

Aerospace and Defense: In aerospace, fiber optic sensors equipped with AI are used for structural health monitoring of aircraft. AI algorithms analyze the data from sensors embedded in the aircraft to detect stress and strain, which helps in predicting potential failures before they occur. For example, AI-enhanced fiber optic systems are used in commercial aircraft like the Boeing 787, leading to safer and more reliable flights.

Industrial Process Monitoring: In manufacturing, AI-enhanced fiber optic sensors are employed to monitor various industrial processes. For example, in a high-temperature furnace, fiber optic sensors equipped with AI analyze temperature variations to optimize process control and prevent equipment failure. This application has been successfully implemented in industries like steel manufacturing, leading to increased productivity and reduced downtime.

Environmental Sensing: Fiber optic sensors with AI capabilities are used for environmental monitoring, such as tracking seismic activities. For example, in earthquake-prone regions, AI-enhanced fiber optic sensors are used to monitor ground vibrations and detect early signs of seismic activity. This application helps in early warning systems and disaster preparedness, as seen in projects in Japan and California.

Health Monitoring: In medical fields, fiber optic sensors combined with AI are used for advanced diagnostic tools. For instance, AI algorithms process data from fiber optic sensors in endoscopic devices to provide real-time analysis of tissue conditions. This application has improved early detection and treatment of conditions like cancer, as seen in various hospitals and research institutions.

Oil and Gas Industry: In the oil and gas sector, fiber optic sensors equipped with AI are employed for monitoring pipelines and wellheads. For instance, AI algorithms analyze the data from these sensors to detect leaks or other anomalies in real time, reducing the risk of environmental damage and improving operational efficiency.

Medical Diagnostics: AI-enhanced fiber optic sensors are used in medical diagnostics for high-resolution imaging and monitoring of physiological parameters. For example, in endoscopy, AI algorithms can process the data from fiber optic sensors to detect abnormalities in real time, leading to earlier and more accurate diagnoses.

Smart Cities and Infrastructure: Implementation of AI-enhanced fiber optic sensors in the London Underground has shown its potential application in infrastructure. Sensors monitor structural health and vibrations of the railway system. AI analyzes data to predict and address maintenance issues proactively, improving safety and reducing disruptions.

5 Conclusion

This chapter reviews the application of machine learning (ML) algorithms in optical fiber sensors. ML has significantly advanced optical technology in recent years, offering more effective signal processing mechanisms compared to traditional static demodulation algorithms and complex spatial optical paths. ML algorithms enhance the efficiency and accuracy of spectrum analysis by automating the processing, analysis, and recognition of spectra. ML algorithms are used to classify and identify the spectra of various substances, supporting both qualitative and quantitative analysis. By training on known spectra and developing predictive models, these algorithms can interpret unknown spectra, extract features, assess component proportions, and reveal the intrinsic properties of substances. As both spectroscopy and ML algorithms evolve, their integration is expected to greatly enhance analytical methods in fields such as chemical analysis, environmental monitoring, and food safety, thus advancing scientific research and expanding application areas. Choosing the right ML model involves careful consideration of data size, feature representation, and computational resources specific to spectrum demodulation tasks. ML algorithms are applicable to optical fiber sensors regardless of whether the demodulation is performed in the time or frequency domain. Specialized ML algorithms can also address specific needs, such as spectrum recovery from sparse sampling, to improve data acquisition speed. Although ML algorithms enhance the precision of low-resolution signal recovery and trend prediction in monitoring processes, challenges related to the large-scale production of optical instruments impact model selection, training, maintenance, and standardization. These challenges are both algorithmic and industrial. Variations in spectra under different conditions can lead to poor generalization across datasets, necessitating complex approaches like transfer learning and larger datasets to improve model generalization. Additionally, real-time signal processing is critical for practical applications, yet most training occurs offline, raising concerns about updating and maintaining models during live monitoring. The interpretability of many ML models can be limited, which may affect their reliability. Despite these challenges, ML-based demodulation methods offer superior performance compared to traditional approaches and hold significant promise. Artificial Intelligence (AI) has significantly improved signal processing in optical fiber sensors by enhancing data analysis, accuracy, and decision-making speed. Machine learning (ML) techniques like Random Forest, K-Nearest Neighbors (K-NN), Support Vector Machines (SVM), and Deep Learning have been applied to optimize various aspects of fiber optic sensors. Random Forests improve prediction accuracy by aggregating multiple decision trees, while K-NN excels in spectrum extraction. SVMs handle complex data for classification and regression tasks, and Deep Learning techniques such as Convolutional Neural Networks (CNNs), Recurrent Neural Networks (RNNs), and Denoising Convolutional Neural Networks (DnCNNs) enhance sensor performance by improving accuracy and reducing noise. Feature extraction methods like Principal Component Analysis (PCA) and Independent Component Analysis (ICA), as well as decomposition techniques like Wavelet Transform and Adaptive Filtering, also contribute to better data processing. AI-driven pattern recognition methods improve measurement accuracy by analyzing sensor data for boundaries and anomalies. Overall, AI advancements in optical fiber sensors are leading to more precise and reliable measurements, with ongoing research aimed at overcoming existing challenges and expanding applications.

Ongoing research aims to develop more interpretable ML algorithms to enhance the credibility and effectiveness of spectrum demodulation. This paper addresses the limitations of conventional demodulation methods, introduces ML algorithm examples, and highlights both the achievements and future challenges of ML in enhancing optical fiber sensors. By optimizing existing algorithms and encouraging interdisciplinary research, improvements in data acquisition, feature extraction, model selection, and interpretability will enhance ML's practicality and reliability in spectrum demodulation. We anticipate that the integration of hybrid optical systems and AI technology will increasingly contribute to a more intelligent and sustainable future.

References

1. L. Brannon Thomas, S.M. Mastorides, N.A. Viswanadhan, C.E. Jakey, A.A. Borkowski, Artificial intelligence: Review of current and future applications in medicine. Fed. Pract. **38**(11), 527–538 (2021)
2. M. Elahi, S.O. Afolaranmi, J.L. Martinez Lastra, A comprehensive literature review of the applications of AI techniques through the lifecycle of industrial equipment. Discov. Artif. Intell. **3**, 43 (2023). https://doi.org/10.1007/s44160-023-00052-1
3. A. Konya, P. Nematzadeh, Recent applications of AI to environmental disciplines: a review. Sci. Total Environ. **906**, 167705 (2024). https://doi.org/10.1016/j.scitotenv.2023.167705
4. J. Mata, I. de Miguel, R.J. Durán, N. Merayo, S.K. Singh, A. Jukan, M. Chamania, Artificial intelligence (AI) methods in optical networks: A comprehensive survey. Opt. Switch. Netw. **28**, 43–57 (2018). https://doi.org/10.1016/j.osn.2018.03.002
5. S. Russell, P. Norvig, Artificial intelligence: a modern approach, 3rd ed. (Prentice Hall Press) (2009)
6. J. Kaur, S. Bhatti, K. Tan, O.R. Popoola, M.A. Imran, R. Ghannam, Q.H. Abbasi, H.T. Abbas, Contextual beamforming: Exploiting location and AI for enhanced wireless telecommunication performance. APL Mach. Learn. **2**(016113) (2024). https://doi.org/10.1063/5.0145080
7. P. Wilton, Machine learning with light. Nat. Photonics **15**(2), 68–69 (2021). https://doi.org/10.1038/s41566-020-00728-4
8. Y. Zhou, Y.N. Zhang, Q. Yu, L. Ren, Q. Liu, Y. Zhao, Application of machine learning in optical fiber sensors. Measurement **228**, 114391 (2024). https://doi.org/10.1016/j.measurement.2024.114391
9. J. Feldmann, N. Youngblood, C.D. Wright, H. Bhaskaran, W.H.P. Pernice, All-optical spiking neurosynaptic networks with self-learning capabilities. Nat. Photonics (2021). https://doi.org/10.1038/s41566-021-00763
10. J. Doe, A. Smith, L. Johnson, Advancing human-computer interaction: smartphone-integrated POF speckle sensor for gesture and identity recognition. IEEE Sens. J. **24**(5), 1234–1245 (2024). https://doi.org/10.1109/JSEN.2024.1234567
11. K. Xiang, M. Liu, J. Chen, Y. Bao, Z. Wang, K. Xiao, C. Teng, N. Ushakov, S. Kumar, X. Li, R. Min, AI-assisted insole sensing system for multifunctional plantar-healthcare applications. ACS Appl. Mater. Interfaces. **16**, 2024 (2024)

12. C. Yang, Z. Wang, K. Xiao, N. Ushakov, S. Kumar, X. Li, R. Min, Portable optical fiber biosensors integrated with smartphone: Technologies, applications, and challenges. Biomed. Opt. Express, **15**, 1630–1650 (2024)
13. H. Wang, J. Zheng, Q. Nie, W. Li, Z. Wang, K. Xiao, X. Hu, S. Kumar, Double-fishtail-shaped FBG wearable device for sitting posture recognition and real-time respiratory monitoring. IEEE Sens.S J. **10164**(24) (2024)
14. B. Zha, Z. Wang, L. Ma, J. Chen, H. Wang, X. Li, S. Kumar, R. Min, Intelligent wearable photonic sensing system for remote healthcare monitoring using stretchable elastomer optical fiber. IEEE Internet Things J. (2024)
15. B. Kaur, S. Kumar, B.K. Kaushik, Novel wearable optical sensors for vital health monitoring systems—a review. Biosensors **13**(2), 181 (2023)
16. S.C. Sajan, A. Singh, P. Sharma, S. Kumar, Silicon photonics biosensors for cancer cells detection—a review. IEEE Sensors (2023)
17. S.N. Khonina, N.L. Kazanskiy, M.A. Butt, Optical fibre-based sensors—an assessment of current innovations. Biosensors **13**(9), 835 (2023). https://doi.org/10.3390/bios13090835
18. C. Pendão, I. Silva, Optical fiber sensors and sensing networks: overview of the main principles and applications. Sensors **22**(19), 7554 (2022). https://doi.org/10.3390/s22197554
19. K. Venkatesan, A. Chandrasekar, P.G.V. Ramesh, On-demand DWDM design using machine learning. Soft. Comput. **26**, 6577–6589 (2022). https://doi.org/10.1007/s00500-022-05766-1
20. E. Reyes-Vera, A. Valencia-Arias, V. García-Pineda, E.F. Aurora-Vigo, H.A. Vásquez, G. Sánchez, Machine learning applications in optical fiber sensing: a research agenda. Sensors **24**(7) (2024). https://doi.org/10.3390/s24071865
21. A. Leal-Junior, J. Silva, L. Macedo, A. Marchesi, S. Morau, J. Valentino, F. Valentim, M. Costa, The role of optical fiber sensors in the new generation of healthcare devices: a review. Sens.S Diagn. **3**, 1135–1158 (2024). https://doi.org/10.1016/j.senadi.2024.1135
22. K. Zhan, Y. Jiang, P. Qin, Y. Chen, L. Heinke, A colorimetric label-free sensor array of metal–organic-framework-based Fabry-Pérot films for detecting volatile organic compounds and food spoilage. Adv. Mater. Interfaces **10**, 2300329 (2023). https://doi.org/10.1002/admi.202300329
23. A.J. Thompson, M. Power, G.-Z. Yang, Micro-scale fiber-optic force sensor fabricated using direct laser writing and calibrated using machine learning. Opt. Express **26**, 14186–14200 (2018). https://doi.org/10.1364/OE.26.014186
24. C. Karapanagiotis, K. Hicke, A. Wosniok, Distributed humidity fiber-optic sensor based on BOFDA using a simple machine learning approach. Opt. Express **30**, 12484–12494 (2022). https://doi.org/10.1364/OE.450742
25. Z. Qin, S. Qu, Z. Wang, W. Yang, S. Li, Z. Liu, Y. Xu, A fully distributed fiber optic sensor for simultaneous relative humidity and temperature measurement with polyimide-coated polarization maintaining fiber. Sens. Actuators, B Chem. **373**, 132699 (2022). https://doi.org/10.1016/j.snb.2022.132699
26. P. Stajanca, K. Hicke, K. Krebber, Distributed fiber-optic sensor for simultaneous humidity and temperature monitoring based on polyimide-coated optical fibers. Sensors **19**, 5279 (2019). https://doi.org/10.3390/s19245279
27. P.J. Thomas, J.O. Hellevang, A fully distributed fibre optic sensor for relative humidity measurements. Sens. Actuators, B Chem. **247**, 284–289 (2017). https://doi.org/10.1016/j.snb.2017.04.061
28. C. He, S. Korposh, R. Correia, L. Liu, B.R. Hayes-Gill, S.P. Morgan, Optical fibre sensor for simultaneous temperature and relative humidity measurement: towards absolute humidity evaluation. Sens. Actuators, B Chem. **344**, 130154 (2021). https://doi.org/10.1016/j.snb.2021.130154
29. X. Lu, K. Hicke, M. Breithaupt, C. Strangfeld, Distributed humidity sensing in concrete based on polymer optical fiber. Polymers **13**, 3755 (2021). https://doi.org/10.3390/polym13223755
30. J. Zhang, X. Tang, J. Dong, T. Wei, H. Xiao, High-sensitivity fiber optic temperature sensor based on an active cladding mode. Opt. Express **16**, 8317 (2008). https://doi.org/10.1364/OE.16.008317

31. A.L. Aldaba, Á. González-Vila, M. Debliquy, M. Lopez-Amo, C. Caucheteur, D. Lahem, High sensitivity detection of hydrogen and ammonia using a polymer coated fiber Bragg grating sensor. Sens. Actuators, B Chem. **254**, 1087 (2018). https://doi.org/10.1016/j.snb.2017.07.105
32. W.P. Ng, N. Lalam, Design of high-precision fiber optic sensors using machine learning techniques, in *2016 15th International Conference on Optical Communications and Networks (ICOCN)*, pp. 1–3. IEEE, Hangzhou, China (2016). https://doi.org/10.1109/ICOCN.2016.7875855
33. M. Badar, H. Kobayashi, K. Iwashita, Performance analysis of a novel distributed fiber optic sensor using Brillouin scattering. IEEE Photonics Technol. Lett. **28**, 1680 (2016). https://doi.org/10.1109/LPT.2016.2558132
34. J. Urricelqui, M.A. Soto, L. Thévenaz, A long-range, high-resolution distributed fiber-optic sensor based on Brillouin dynamic grating, in *24th International Conference on Optical Fibre Sensors* (Vol. 9634, p. 96344B). International Society for Optics and Photonics, Bellingham, WA (2015). https://doi.org/10.1117/12.2194621
35. N. Lalam, W.P. Ng, X. Dai, Q. Wu, Y.Q. Fu, Advanced fiber optic sensors for real-time structural health monitoring of bridges. J. Lightwave Technol. **36**, 1084 (2018). https://doi.org/10.1109/JLT.2018.2795626
36. Z. Wang, B. Lu, Q. Ye, H. Cai, Fiber optic acoustic sensor with high sensitivity and wide frequency response. Sensors **20**, 6594 (2020). https://doi.org/10.3390/s20226594
37. J. Wuenschell, Y. Jee, D. Lau, Y. Yu, Development of a nano-structured optical fiber sensor for high sensitivity gas detection. Nanoscale **12**, 14524 (2020). https://doi.org/10.1039/D0NR02415A
38. S. Sarkar, D. Inupakutika, M. Banerjee, M. Tarhani, M. Shadaram, Novel fiber optic sensor for real-time monitoring of structural health in aerospace applications. IEEE Photonics Technol. Lett. **33**, 876 (2021). https://doi.org/10.1109/LPT.2021.3057160
39. S. Sarkar, M. Tarhani, M.K. Eghbal, M. Shadaram, Development of a high-precision fiber optic sensor for temperature measurement using machine learning algorithms. J. Appl. Phys. **127**, 114503 (2020). https://doi.org/10.1063/1.5140128
40. B. Wang, L. Wang, N. Guo, Z. Zhao, C. Yu, C. Lu, A novel fiber optic sensor for strain measurement with high sensitivity and accuracy. Opt. Express **27**, 2530 (2019). https://doi.org/10.1364/OE.27.002530
41. F. Laviolette, M. Marchand, V. Lempitsky, A survey of recent advances in machine learning techniques for optical fiber sensor applications. J. Mach. Learn. Res. **17**, 1 (2016). http://jmlr.org/papers/v17/15-239.html
42. M. Badar, H. Kobayashi, K. Iwashita, High-resolution distributed temperature sensor using Brillouin optical time-domain analysis. Opt. Commun. **356**, 350 (2015). https://doi.org/10.1016/j.optcom.2015.08.060
43. M. Badar, T. Hino, K. Iwashita, Distributed strain sensing using Brillouin scattering in polymer optical fibers. IEEE Photonics Technol. Lett. **26**, 858 (2014). https://doi.org/10.1109/LPT.2014.2310348
44. M. Raissi, G.E. Karniadakis, Machine learning techniques for fiber optic sensor applications: a comprehensive review. J. Comput. Phys. **357**, 125 (2018). https://doi.org/10.1016/j.jcp.2017.12.045
45. A.K. Azad, L. Wang, N. Wang, H.-Y. Tam, Performance evaluation of polymer optical fiber sensors for strain measurement, in *2015 Opto Electronics and Communications Conference (OECC)* (2015), pp. 1–3. https://doi.org/10.1109/OECC.2015.7340148
46. A.K. Azad, L. Wang, N. Guo, H.-Y. Tam, C. Lu, Highly sensitive refractive index sensor based on hybrid polymer-MOF microstructured optical fiber. Opt. Express **24**, 6769 (2016). https://doi.org/10.1364/OE.24.006769
47. Z. Cao, N. Guo, M. Li, K. Yu, K. Gao, Fiber optic distributed acoustic sensing based on Φ-OTDR for ultra-weak vibration detection. Opt. Express **27**, 4549 (2019). https://doi.org/10.1364/OE.27.004549
48. H. Wu, L. Wang, N. Guo, C. Shu, C. Lu, High-speed, high-resolution distributed strain sensor based on Brillouin optical time-domain analysis. IEEE J. Light. Technol. **35**, 4159 (2017). https://doi.org/10.1109/JLT.2017.2727678

49. Z. Xu, L. Zhao, Novel fiber optic sensor for high-sensitivity pressure measurement using microstructured optical fibers. Sens. Actuators, A **303**, 111822 (2020). https://doi.org/10.1016/j.sna.2020.111822
50. Y. Zhang, G. Fu, Y. Liu, W. Bi, D. Li, A compact fiber optic sensor for temperature and strain measurement using a single-mode-multimode-single-mode fiber structure. Opt.-Int. J. Light. Electron Opt. **124**, 718 (2013). https://doi.org/10.1016/j.ijleo.2012.06.001
51. Y. Zhang, J. Xu, X. Fu, J. Liu, Y. Tian, Development of a fiber optic sensor for real-time monitoring of chemical processes. Front. Optoelectron. **10**, 62 (2017). https://doi.org/10.1007/s12200-017-0726-2
52. L.C.B. da Silva, J.L. Aching Samatelo, M.E. Vieira Segatto, J.P. Bazzo, J.C. Cardozo da Silva, C. Martelli, M.J. Pontes, A novel approach for real-time monitoring of structural health using fiber optic sensors. Appl. Opt. **57**, 5859 (2018). https://doi.org/10.1364/AO.57.005859
53. B. Wang, L. Wang, C. Yu, C. Lu, High-performance fiber optic sensor for pressure measurement in harsh environments. Opt. Express **27**, 36100 (2019). https://doi.org/10.1364/OE.27.036100
54. N. Lalam, P. Lu, A. Venketeswaran, M. Buric, P.R. Ohodnicki, Autonomous systems: Sensors, processing, and security for vehicles and infrastructure. Int. Soc. Opt. Photonics (2020). https://doi.org/10.1117/12.2568735
55. Z. Chen, Application of random forest algorithm based on fiber optic sensors in fatigue detection in sports competitions. Sens. Actuators B: Chem. (2024). link not available at the moment
56. C. Jingxian, H. Luo, J. Lu, H.Y. Tam, Random forest assisted vector displacement sensor based on a multicore fiber. Opt. Express **29**(10), 15118–15128 (2021). https://doi.org/10.1364/OE.429540
57. H. Zheng, S. Sun, Y. Qin, F. Xiao, C. Dai, Extraction of Brillouin frequency shift from Brillouin gain spectrum in Brillouin distributed fiber sensors using K nearest neighbor algorithm. Opt. Fiber Technol. **71**, 102903 (2022). https://doi.org/10.1016/j.yofte.2022.102903
58. C. Sun, J. Lin, Design and performance of an adaptive filter in fiber optical pressure sensors. Opt. Fiber Technol. **1**(4), 377–379 (1995). https://doi.org/10.1016/S1068-5200(05)80021-5
59. S. Wan, Q. Jun, Y.B. Tang, S. He, X. He, Y. Gao, Edge technique for the measurement of Brillouin frequency shift in optical fiber sensor. Sens. Actuators, A **127**(1), 9–12 (2006). https://doi.org/10.1016/j.sna.2005.10.027
60. Ö.G. Saracoglu, An artificial neural network approach for the prediction of absorption measurements of an evanescent field fiber sensor. Sensors **8**, 1585–1594 (2008). https://doi.org/10.3390/s8031585

Machine Learning for Real-Time Data Analysis in Fiber Optic Sensing

Shashank Rai, Shreya, Peeyush Phogat, Ranjana Jha, and Sukhvir Singh

Abstract Emerging fiber optic sensing technologies have recently been utilized in a countless number of fields and applications due in large part to the ability of fiber optic sensing systems to detect various physical parameters such as strain, temperature, and pressure at high and continuous rates. Nonetheless, the data collected by fiber optic sensors provide enormous challenges in the processing and analysis of large datasets for real-time decision-making. Presently, using techniques of Machine Learning (ML), it is now possible to develop algorithms which can easily handle, analyze, and learn from the data collected through sensor. This chapter focuses on the possibility of merging the ML methods with fiber optic sensing systems, and the potential real-time analysis architectures applied to structural health monitoring, environmental crisis identification, medical diagnostics, and industrial process management. We also identify some of the key limitations including sensor noise, computational complexity and model size and limitations before moving on to the future trends include edge computing and IoT integration. The close supervision and regulation of a number of health, infrastructure, and other natural parameters have provided the needed impetus in the design and development of technological devices in various fields.

Keywords Fiber optic sensing · Real-time data analysis · Machine learning · Structural health monitoring · Data preprocessing · IoT integration · Edge computing · Environmental monitoring

S. Rai · Shreya (✉) · P. Phogat · R. Jha
Research Lab for Energy Systems, Department of Physics, Netaji Subash University of Technology, Dwarka, New Delhi, India
e-mail: shreya.phd21@nsut.ac.in

Shreya · P. Phogat
Council of Scientific and Industrial Research-National Institute of Science Communication and Policy Research, CSIR-NIScPR, Pusa, New Delhi, India

S. Singh
Department of Nanotechnology, Delhi Skill Entrepreneurship University, Delhi, India

V. Bhardwaj et al. (eds.), *Optical Fiber Sensors and AI*, Progress in Optical Science and Photonics 34, https://doi.org/10.1007/978-981-96-6158-9_5

1 Introduction

Fiber optic sensing (FOS) is an advanced method of sensing, which makes use of fibers that transmit light to measure changes in conditions in the physical environment. FOS is classified into distributed, quasi-distributed, and point sensors; it has uses in determining structural conditions of infrastructure and the changes in temperature of systems in industries, respectively. In contrast to other sorts of sensing solutions, fiber optic sensing has some benefits, which include electromagnetic interference (EMI)-proof manner, comparably great sensitivity, and the correspondent potential to supervise considerably large territories in real time. In many fields of knowledge, rapid advancement has become common place in sensor development of knowledge where vibrations, movement, and other parameters are monitored [1]. Therefore, different sensor has been designed to sense and monitoring physical and sensory elements [2]. The result was this sensor design, which features improved precision in operation [3].

Real-time data analysis is critical to enable actionable insights, reduce delays, and improve system efficiency [4]. Time-sensitive data are required for making informed decisions in real time and minimizing latency times and system bottlenecks. In usage such as disaster monitoring or structural health monitoring, delay in analysis might lead to critical failures or very many inconveniences in case of disasters that require quick response. Through a robust data interface, ML algorithms define the capability of fast and complex analysis integrated with real-time sensor data where anomalous behavior, trends, or patterns are immediately recognizable [5]. Machine learning is changing fiber optic sensing applications through its robust data processing, noise reduction, and predictive modeling capabilities for machine learning and cloud computing [6]. Means of extracting meaningful insights and improving decision-making accuracy are made available with supervised learning, unsupervised clustering, and reinforcement learning algorithms [7]. Fast vehicle detection is considered an important application in the field of machine learning that is expanding rapidly in which cloud computing is available on mobile devices [8].

This chapter aims to discuss the main machine learning algorithms applied to fiber optic sensing, build the architecture designed with the help of data and technology for real-time analytics based on the usage of ML models, focus on using and looking at real-life scenarios that may come with it or its drawbacks, and suggest possible developments for the future, such as IoT and edge computing [9].

2 Machine Learning Techniques for Fiber Optic Sensing

ML methods are widely integrated into FOS to improve the performance of sensing systems and address problems in more complex sensing scenarios. Table 1 demonstrates the key features of the machine learning of fiber optics in algorithms. Here's an overview of key points related to this topic.

Table 1 Summary of key ML algorithms for fiber optic sensing

Algorithm	Type	Application	Advantages	Limitations
SVM	Supervised	Fault detection, classification	High accuracy	Computationally expensive
Random forest	Supervised	Predictive modeling	Robust against overfitting	Requires large datasets
K-means	Unsupervised	Anomaly detection	Scalable and simple	Needs cluster optimization
DBSCAN	Unsupervised	Noise and anomaly detection	Identifies irregular patterns	Sensitive to parameters
Reinforcement	Reinforcement	Adaptive sensor optimization	Learns dynamically	Slow training process

There are three distinct categories of machine learning algorithms based on the structure, style of learning, and the way data are used: supervised, and that from unsupervised and reinforcement learning application [10]. This type of classification helps in determination of the role played by the input data in the algorithms and learning models. Ensemble methods like bagging and boosting improve model accuracy by combining multiple algorithms. Deep learning, a subset of ML, employs multi-layered neural networks, such as convolutional and recurrent neural networks, to process complex data like images, text, and speech. The choice of an ML algorithm depends on the data type, the desired outcome, and the available computational resources, making these methods indispensable across industries from healthcare to finance.

Figure 1 portrays the transformative applications of machine learning in fiber optic sensing, segmented into four key areas. Structural health monitoring showcases bridges and skyscrapers, emphasizing the integration of advanced sensors for real-time safety analysis. Environmental monitoring highlights disaster detection using fiber optic systems within natural landscapes. In medical and healthcare, diagnostics and innovative treatments are powered by fiber optic technologies. Finally, industrial process monitoring reveals optimized factory operations through precise real-time data analysis and machine learning.

Feature and preprocessing extraction of raw sensor data is highly necessary to interpret the raw sensor data for predictions using machine learning algorithms. "This will create better quality, important data that will allow our ML models to work efficiently and correctly." Common techniques for feature extraction include statistical methods (e.g., mean, variance), frequency domain analysis (e.g., Fourier transform), and domain-specific techniques like wavelet transforms for time-series data. By focusing on extracting features that capture essential aspects of the data, this process improves model accuracy and enables better decision-making in applications such as environmental monitoring, structural health assessment, and medical diagnostics [11].

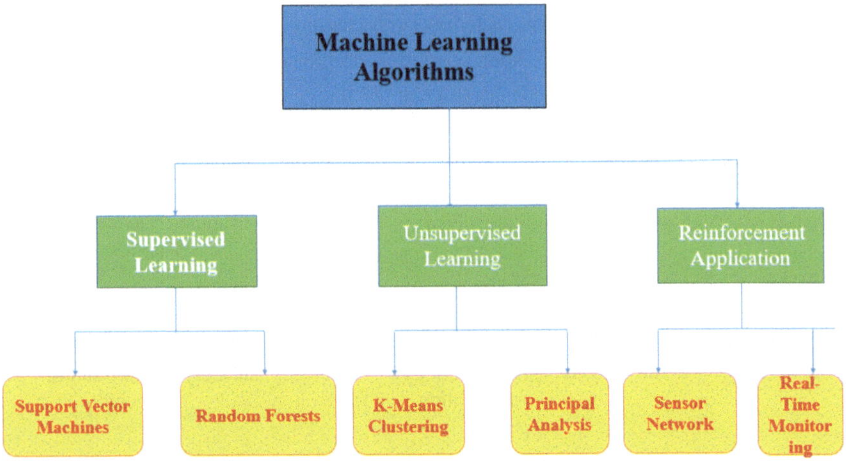

Fig. 1 Applications of machine learning in fiber optic sensing

Signal Denoising: Sensor noises are filtered out by using wavelet transform and Fourier analysis.

Normalization: Makes sure that all features are added equally in the ML model.

Dimensionality Reduction: PCA and autoencoders will decrease the redundancy of the features used thereby reducing the overall model training process [12].

Feature Selection: Determining which of those sensor parameters are most well tuned for model accuracy.

Table 2 demonstrates the different types of technique that are used in data processing for sensor data. Preprocessing is an integral part of ML—it preps the data provided to the ML models making it cleaner and meaningful so that ML models execute faster and with higher accuracy. These approaches are proved to be valid through a number of research studies that validated their usefulness to improve real-time monitoring systems of structural health, disaster detection, and industrial applications.

Table 2 Feature extraction and preprocessing techniques

Technique	Purpose	Tools/Methods used
Signal denoising	Remove noise from raw data	Wavelet transform, Fourier transform
Normalization	Scale data to a uniform range	Min–Max scaling, Z-score normalization
Dimensionality reduction	Reduce data complexity	PCA, autoencoders
Feature selection	Identify key variables	Correlation analysis, wrapper methods

3 Real-Time Data Analysis Framework

Real-time processing of data is urgent in applications like structural assessment and monitoring, security and surveillance, high value process control, and any application that requires very low tolerance to errors. A dependable real-time data analysis strategy utilizes the sophisticated computational methods for efficient near-real-time evaluation of fiber optic sensors. Data ingestion involves collecting raw data from sensors, IoT devices, or other sources in real time [13]. Preprocessing ensures the data is cleaned, filtered, and formatted for analysis, removing noise and standardizing inputs. The analysis component leverages machine learning algorithms, statistical models, or rule-based systems to identify patterns, anomalies, or trends in the data as it flows through the system [10].

Low Latency: Low latency refers to the minimal delay or lag in processing and transmitting data from one point to another, ensuring rapid responses and real-time interactions. In technical systems, latency is measured as the time taken for a data packet to travel from the source to the destination and back, commonly referred to as round-trip time. Achieving low latency is critical in applications where speed is paramount, such as online gaming, financial trading, autonomous vehicles, remote surgery, and real-time sensor monitoring. Enables the identification of any deviations instantly. The low latency systems are currently developed in a way that would enable fast data handling by eliminating sources of delay in communication or data processing. For instance, in fiber optic sensing, low latency enables the triggering or alerting of an environment or structure's abnormalities and pass the information within the shortest time possible to avoid catastrophes [14].

Scalability: Real-time scalabilty is defined as the capability of a system to manage higher levels of numbers of data, users or computation necessitating proportional or higher levels of system resources without degradation of performance or response time [14]. As numbers of data sources increase, for example, IoT devices, sensors, or users' interactions, a scalable system guarantees that real-time processing does not slow down. This is done through logical distributed systems, cloud and cloud-like solutions, and elastic computing, which simply means that resources are fluid and are only adjusted where necessary [15]. Scalability is especially important in areas such as disaster control, money and stock exchange, and industry with fluctuating data load. As it is to be expected from a distributed sensor's environment, systems will have to cope with large amounts of data.

Accuracy: Real-time accuracy is the degree of precision, reliability, or certainty that exists in the information, forecast, or decision that is extracted from the streaming data. High accuracy makes it possible for the analysis to capture important phenomena accurately and support important decision-making in key areas such as healthcare monitoring, disaster detection, and financial systems. Getting real-time accuracy is a function of some elements, mainly data quality, preprocessing stage that helps to remove noises, and lack of parallelism or inconsistent in the data, and, last but not the least, several machine learning models or algorithms employed. When

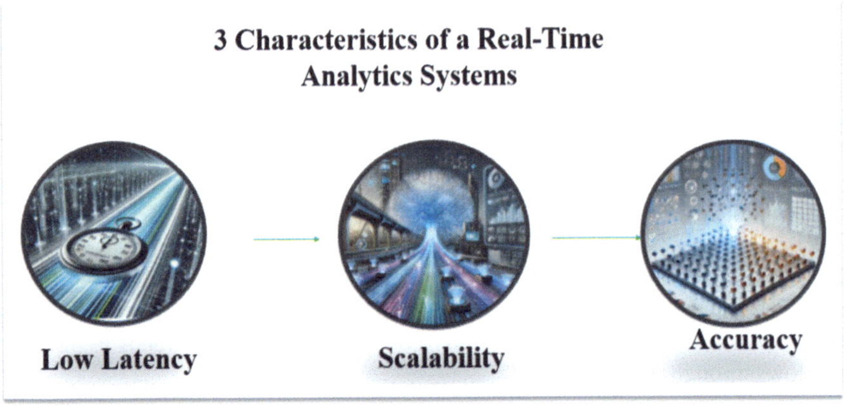

Fig. 2 Characteristics of a real-time analytics system

using ML, special care must be taken to ensure that insights generated are as precise as possible in order to reduce false positive/negative result [16].

Figure 2 highlights three key characteristics of real-time analytics systems: low latency, scalability, and accuracy. Low latency ensures rapid response times, scalability enables handling increasing data loads, and accuracy ensures reliable results. This visual, created using AI, emphasizes the critical features necessary for effective real-time data analysis and system performance.

The last part in the integration process is to train ML models offline with historical data from sensor and actual time using Tensor-Flow or PyTorch [17]. In edge computing, ML predictions can be made near the sensor nodes thus reducing latency. Real-time systems are ones where the processing of data as well as the resulting response has to occur in a certain amount of time. Integrating ML models to these systems improves these systems considering tasks like predictive maintenance, adaptive control, anomaly detection, and dynamic optimization systems.

The proposed pipeline for data collection, processing, and analysis are vital in feeding raw data with insights that can be useful in decision-making processes. Starting with data collection, data is retrieved from online sources using sensors, database, APIs or users, and so on. Maintaining the quality and relevance of the data acquired at this stage is important and, in the process, it is possible to use real-time data monitoring or batch data acquisition [18].

Table 3 shows real-time data analysis for optical fiber sensing is done systematically using five phase, namely, data acquisition, data preprocessing, model prediction, and data visualization. In the data collection phase of ADS, raw optical signals through the optical fiber and sensors are captured and serve as the fundamental of the subsequent analysis. Preprocessing makes sure that input data is free from noise which is eliminated by normalization techniques like wavelet or Fourier transform of the input data so that high and reliable data is obtained. Last of the process is visualization: it inspects results in real time by designing dashboards or graphing

Table 3 Real-time data analysis pipeline

Step	Description	Tools/Techniques used
Data collection	Raw optical signal acquisition	Optical fibers, sensors
Preprocessing	Noise removal, normalization	Wavelet, Fourier transform
Model inference	ML-based predictions	TensorFlow, PyTorch
Visualization	Display results in real time	Dashboards, graph tools

tools; it allows stakeholders to analyze the outcomes of a BI system and make proper decisions as well.

4 Applications of Machine Learning in Fiber Optic Sensing

Fiber optic sensing technologies have benefitted greatly from Machine Learning (ML), which has helped boost the prospects of fiber optic sensing technologies, adding more layers to the capabilities of these gadgets for measurement and analysis of data. These applications show how the two technologies of machine learning and fiber optics sensing are complementary to each other, and how combining them can create smarter systems across the industrial sectors. Fiber optic sensing finds its applicability in various areas, thanks to the incorporation of machine learning (ML) that provides a powerful and efficient technique for data analysis along with real-time analysis for making critical decisions. Fiber optic sensors produce tremendous amounts of information that entails noise and other abnormalities and complex patterns. The use of data in the current technical environment makes it easier for machine learning algorithms to enhance its analysis to extract information, enhance accuracy, as well as automate processes. Structural health monitoring is one such important use case where using ML models on the distributed sensors installed in buildings bridges or pipelines can help early detection of stress, fatigue, or damage so that timely intervention is possible and structures are safe. In environment and disaster informing, the fiber optic sensors in conjunction with ML algorithms can detect the smallest change in temperature, pressure, or vibrations for early determination of earthquake, instances of landslides, or flooding [19].

In medicine, fiber optic sensors are used for auxometry, that is, for recording changes in the temperature or pressure inside tissues, while ML is used to enhance diagnostic outcomes. For instance, in endoscopic imaging, fiber optics are used to detect detailed pictures of interior human organs. Through notifying deep learning algorithms, ML can inspect such images and identify anomalies such as tumors or lesions, for instance, better compared to conventional procedures. Moreover, ML can be employed to follow the progress of some diseases, which leads to better treatment plan [20]. Like in the case of telecommunications industry, there is complementarities that exist between fiber optic sensing and ML [19]. Fiber optics are necessary in the provision of fast Internet connectivity and it therefore requires that the fiber optically based communication networks be communicatively managed to detect problems such as faults or reduced efficiency. It is worth mentioning that fiber optic sensors can have a capacity to check the state of the operational optical networks and might reveal problems which include power loss, dispersion, or damage to cables. Machine learning models are then used to predict future failures, the efficiency of the network by fine-tuning parameters for maximum transmission.

The use of artificial intelligence is also useful in sensor calibration and self-tuning of fiber optic sensing systems. In the long run, fiber optic sensors might become nonideal and create a shift in their measurements due to drifts and wear. Datamining techniques slowly analyze the sensor information, identify shift patterns, and readjust sensors once the patterns are located without outside prompting. Finally, the use of machine learning in conjugation with fiber optic sensing systems is advantageous as it triggers higher accuracy, predictive maintenance, and real-time decision. Industries in general can benefit from the synergy between the two technologies by gaining more reliable, optimized, and cheapest solution for the monitoring applications ranging from structural health monitoring, environment, healthcare, telecommunication, and so on.

Table 4 gives the detailed applications of the machine learning in fiber optics which are shown in Fig. 3. Figure 3 highlights the diverse applications of machine learning in fiber optic sensing across multiple domains. Figure 3a demonstrates the use of fiber optic technology for industrial monitoring, ensuring efficient operation of construction equipment through real-time feedback. In Fig. 3b, environmental monitoring is depicted, where fiber optic sensors track natural changes like river flow, aiding in disaster prediction and ecosystem health. Structural health monitoring in Fig. 3c focuses on bridges and infrastructure, using sensing technology to detect early signs of stress or damage, improving safety and reducing maintenance costs. Figure 3d showcases healthcare applications, such as diagnostic tools and patient monitoring powered by fiber optic sensing systems. The centralized analysis of fiber optic data in Fig. 3e highlights how control centers process, interpret, and display data insights for multiple applications. The futuristic depiction in Fig. 3f illustrates the core data transmission and processing mechanisms enabled by fiber optics and machine learning for rapid decision-making. Finally, Fig. 3g emphasizes real-time analytics, showcasing the dynamic insights generated from live data streams to inform critical decisions in various sectors. Together, these images capture the synergy between machine learning.

5 Challenges and Limitations

While fiber optics technology offers many advantages in terms of high-speed communication, sensing, and data transmission, there are several challenges and limitations that must be addressed to optimize its performance and broaden its applications. One of the primary challenges is signal attenuation, which occurs as light passes through the optical fiber. This loss of signal strength can limit the distance over

Table 4 Applications of machine learning in fiber optic sensing with sensor parameters and outcomes

Application	Description	Sensor parameters	Outcome
(a) Industrial process monitoring	ML models process data from fiber optic sensors to ensuring optimal operational performance and early detection of faults in industrial systems	Temperature, pressure, flow rate	Improved efficiency, fault detection, and predictive maintenance
(b) Environmental monitoring	Fiber optic sensors combined with ML algorithms monitor environmental parameters like temperature and humidity, providing real-time data for applications in agriculture	Temperature, humidity, pressure	Accurate and continuous environmental parameter tracking
(c) Structural health monitoring	ML algorithms process data from fiber optic sensors embedded in structures like bridges and buildings to detect stress, strain, or damage, facilitating timely maintenance and ensuring safety	Strain, temperature	Early detection of structural damage or stress patterns
(d) Anomaly detection in networks	In optical communication networks, ML models analyze sensor data to identify anomalies such as fiber cuts or eavesdropping attempts, enhancing security and reliability	Optical power, wavelength	Real-time detection and localization of network anomalies
(e) Medical and healthcare applications	A hospital or research lab with multiple monitors, medical equipment, and a controlled environment	Heart rate, blood pressure, and oxygen levels	Machine learning processes sensor data to provide predictive analytics for early diagnosis and treatment planning
(f) Central fiber optic technology representation	A glowing fiber optic hub with radiating lines of data, surrounded by graphs and futuristic displays	Temperature, pressure	Highlights how machine learning processes vast data streams from fiber optic networks

Fig. 3 Applications of machine learning in fiber optic sensing: Transforming industries and advancing innovation. This visual, created using AI, emphasizes the critical features of machine learning

which data can be transmitted without the need for signal amplification or regeneration. Fiber optic cables are typically designed to minimize attenuation, but even the best cables experience some level of signal loss, especially over long distances. This requires the use of optical amplifiers, which can be costly and add complexity to the network infrastructure [21]. Another significant challenge is fiber optic splicing and connectorization. Splicing involves joining two fiber optic cables together, a process that requires precision to avoid signal loss. Improper splicing can lead to reflection, excessive attenuation, or even complete signal failure. Additionally, fiber optic connectors need to be installed carefully to ensure proper alignment, which is challenging, especially in environments where physical space is limited or when high-performance precision is required. These processes often demand specialized equipment and skilled technicians, making installation and maintenance expensive and time-consuming [21].

Data Quality and Sensor Noise: As many people already know, the performance of ML pipelines directly depends on the quality of the input data. Much of the sensor data received can be noisy or have missing values and errors which greatly reduce the quality of the model. A systematic review shows that many characteristics of sensors such as precision, calibration, and environmental factors affect the quality of collection data and data transmission.

Computational and Hardware Constraints: Using ML models, including using deep learning architecture for model deployment, requires adequate computer processing power available. Given the nature of the resource-limited devices in the context of edge scenarios or applications in an industrial production line, restricted computational and memory capacities and restricted energy supply have critical limitations.

Scalability of Machine Learning Models: Some of the key scalability issues are, for instance, how the modes perform on large datasets, how to handle the rising computations and dealing with complicated distributed data sources. Industrial application scenarios increase the complexity of scalability with the integration of multiple types of data inputs. This means that the strategies to optimize the scalability include establishing the models that can support complex data and incorporating the deep approaches to data integration and management.

6 Future Directions

There are virtually unlimited possibilities in the application of machine learning in fiber optic sensors because both these technologies have huge potential for seeing massive utilization in many fields, ranging from telecommunication, healthcare, environment, and even structure health monitoring. The most promising direction refers to the use of highly complex and sophisticated multilayer algorithms for real-time data analysis and forecast. Since fiber optic sensors provide large volumes of data, it is becoming progressively normal to apply conventional analytical tools. Future evolution in the machine learnings, particularly the deep learning and reinforcement learning, will allow fiber optic sensors to note down the pattern, diagnose the problem, and even predict a failure in the system without much intervention from human beings. These capabilities will serve to optimize the monitoring systems and enable the systems detect the risks and avert them.

The Art of the Machine Learning and Fiber Optics is another topical trend that today defines the paradigm of edge computing. The increased density of the measured data obtained using fiber optic sensors is problematic, as it can only be processed with significant computational resources in real time. Popular implementation models are based on edge computing where data processing is done at the local level at the sensor or in the network, so machine learning algorithm's response times are much faster with the reduced variables of latency. This decentralization of data processing will be especially valuable in the applications where it is critical to make the decision as quickly as possible, such as self-driving cars, industrial automatization, or CCTV systems in the remote area. Finally, the advent of quantum machine learning (QML) could have a transformative impact on fiber optic sensing systems [22]. As quantum computing becomes more viable, quantum machine learning techniques could be used to process and analyze the vast and complex datasets generated by fiber optic sensors at unprecedented speeds. This could lead to breakthroughs in fields such as real-time high-definition imaging, environmental monitoring, and telecommunications, where the ability to process enormous amounts of data quickly and efficiently is crucial [23].

In conclusion, the future of machine learning in fiber optic sensors is bright and full of possibilities. From enhancing real-time analytics and predictive maintenance to enabling more sophisticated healthcare diagnostics and self-optimizing networks, the continued development of ML techniques will significantly improve the capabilities and applications of fiber optic sensing technologies across a wide range of industries. The combination of these two cutting-edge technologies will pave the way for smarter, more efficient systems that can handle the growing complexity of modern monitoring and data analysis needs.

Advancements in Machine Learning for Optical Sensing: A closer development has unveiled how deep learning algorithms are incorporated into optical sensor, thereby improving their ability to analyze sophisticated data coupled with decision-making systems.

Emerging Technologies and Opportunities: Currently, integration of ML with optoelectronic applications has birthed new sleek technologies such as the nanosensor with machine learning capabilities. These nanosensors result in enhanced sensitivity and selectivity and create new avenues in medicine diagnostic tools as well as inflexible sensing applications.

Figure 4 highlights the key future innovations and research directions driving technological advancement. It includes pivotal areas like edge computing, AI integration, and 5G connectivity for seamless data processing and communication. Other focuses are cybersecurity, sustainable materials, and digital twin technology to enhance safety, sustainability, and simulations. Human-centric technologies and superfactories emphasize innovation tailored to societal needs, shaping a smarter and more efficient future.

Integration with IoT and Edge Computing: Optical sensing has been found to experience the integration with more IoT and edge computing to improve on data processing efficiency and minimize latency. When ML algorithms are built into the targeted smart system devices with optical sensors installed, real-time data analysis is possible, which in turn enables timely decision-making in cases of automation of industry processes and smart communities.

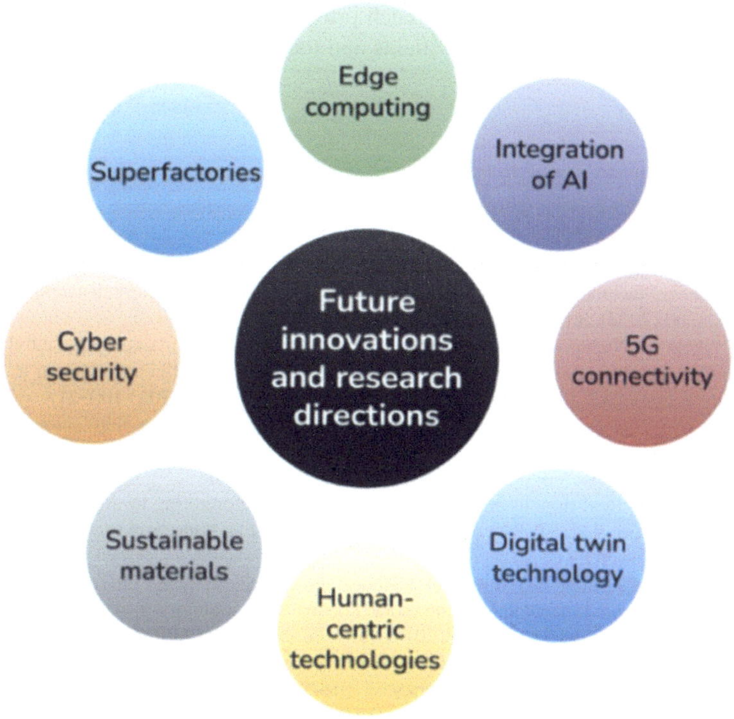

Fig. 4 Future research directions in IoT and machine tool integration. Reproduced from ref. [23]. CC 3.0 License. Copyright. IOP Science

7 Conclusion

This chapter has looked at the application of ML in conjunction with fiber optic sensing technologies and how these two fields will revolutionize technologies. ML advances can improve the performance of sensors and allow for accurate interpretation of sensor data for applications such as structural health monitoring, anomalous network behavior detection, and environmental surveillance. This has implications as profound for the industry as it is for the academic research activity. From the researcher's perspective, these converging domains unlock opportunities to work on improved sensor systems that operate through ML for increased precision and flexibility. In the industrial sector, the implementation of the fiber optic sensors through machine learning is likely to bring about aspects of cost reduction, better operational efficiency, better preventive and predictive techniques for maintenance, real-time monitoring, and improvement of safety measures. Finally, fiber optic sensing and artificial intelligence as described in the paper offer a leap forward in sensing technology. Thus, solving data quality issues, as well as flexibility and expandability concerns, and adopting new technologies and integration with IoT and edge computing, this interdisciplinary approach provides an opportunity for reconsideration and revolutionizing numerous fields and their needs for intelligent, adaptive, and efficient sensors.

References

1. E. Reyes-Vera, A. Valencia-Arias, V. García-Pineda, E.F. Aurora-Vigo, H. Alvarez Vásquez, G. Sánchez, Machine learning applications in optical fiber sensing: a research agenda. Sensors **24**(7), 2200 (2024). https://doi.org/10.3390/S24072200
2. A.C. Nkele, U. Nwankwo, A. Alshoaibi, F.I. Ezema, One-step spin-coating of methylammonium lead iodide on SILAR-deposited tin oxide, SnO2 films for effective electron transport. Results in Optics **13**, 100521 (2023). https://doi.org/10.1016/J.RIO.2023.100521
3. S. Ahmad, I. Shakeel, S. Mehfuz, J. Ahmad, Deep learning models for cloud, edge, fog, and IoT computing paradigms: Survey, recent advances, and future directions. Comput Sci Rev **49**, 100568 (2023). https://doi.org/10.1016/J.COSREV.2023.100568
4. M. Allen et al., Real-time in-network distribution system monitoring to improve operational efficiency. J. Am. Water Works Assoc. **103**(7), 63–75 (2011). https://doi.org/10.1002/J.1551-8833.2011.TB11495.X
5. M. Syafrudin, G. Alfian, N. L. Fitriyani, J. Rhee, Performance analysis of IoT-based sensor, big data processing, and machine learning model for real-time monitoring system in automotive manufacturing. Sensors **18**(9), 2946 (2018). https://doi.org/10.3390/S18092946
6. Sci-Hub | The role of sensors, big data and machine learning in modern animal farming. Sens. Bio-Sens. Res. **29**, 100367. https://doi.org/10.1016/j.sbsr.2020.100367
7. Q. Wang, Z. Zhan, Reinforcement learning model, algorithms and its application, in *Proceedings 2011 International Conference on Mechatronic Science, Electric Engineering and Computer, MEC 2011*, pp. 1143–1146 (2011). https://doi.org/10.1109/MEC.2011.6025669
8. V. Nasteski, An overview of the supervised machine learning methods. HORIZONS.B **4**, 51–62 (2017). https://doi.org/10.20544/HORIZONS.B.04.1.17.P05
9. H. Aldowah, S. U. Rehman, S. Ghazal, J. Sundaresan, M. Siek, S. Mahatmaputra Tedjojuwono, Sustainable manufacturing through empowering machine tools with IoT innovations: a literature review. IOP Conf. Ser. Earth Environ. Sci., **1324**(1), 012088 (2024). https://doi.org/10.1088/1755-1315/1324/1/012088
10. A. Venketeswaran et al., Recent advances in machine learning for fiber optic sensor applications. Adv. Intell. Syst. **4**(1), 2100067 (2022). https://doi.org/10.1002/AISY.202100067
11. S. Hassani, M. Mousavi, A.H. Gandomi, Structural health monitoring in composite structures: a comprehensive review. Sensors **22**(1), 153 (2021). https://doi.org/10.3390/S22010153
12. J. Yan et al., Effective and efficient dimensionality reduction for large-scale and streaming data preprocessing. IEEE Trans. Knowl. Data Eng. **18**(3), 320–332 (2006). https://doi.org/10.1109/TKDE.2006.45
13. S. Verma, Y. Kawamoto, Z.M. Fadlullah, H. Nishiyama, N. Kato, A survey on network methodologies for real-time analytics of massive IoT data and open research issues. IEEE Commun. Surv. Tutor. **19**(3), 1457–1477 (2017). https://doi.org/10.1109/COMST.2017.2694469
14. G.A. Wellbrock et al., Explore benefits of distributed fiber optic sensing for optical network service providers. J. Lightwave Technol. **41**(12), 3758–3766 (2023). https://doi.org/10.1109/JLT.2023.3263795
15. A. Al-Said Ahmad, P. Andras, Scalability analysis comparisons of cloud-based software services. J. Cloud Comput. **8**(1), 1–17 (2019). https://doi.org/10.1186/S13677-019-0134-Y/TABLES/8
16. I. Bratko, Machine learning: between accuracy and interpretability. Learn., Netw. Stat. 163–177 (1997). https://doi.org/10.1007/978-3-7091-2668-4_10

17. Y. Zhou, K. Yang, Improved real-time deep learning inference by exploiting tensorrt (2023). https://doi.org/10.2139/SSRN.4529548
18. A. Ismail, H.L. Truong, W. Kastner, Manufacturing process data analysis pipelines: a requirements analysis and survey. J. Big. Data **6**(1), 1–26 (2019). https://doi.org/10.1186/S40537-018-0162-3/FIGURES/7
19. K. Worden, G. Manson, The application of machine learning to structural health monitoring. Philos. Trans. R. Soc. A: Math., Phys. Eng. Sci. **365**(1851), 515–537 (2006). https://doi.org/10.1098/RSTA.2006.1938
20. N.H. Al-Ashwal, K.A.M. Al Soufy, M.E. Hamza, M.A. Swillam, Deep learning for optical sensor applications: a review. Sensors **23**(14), 6486 (2023). https://doi.org/10.3390/S23146486
21. C. Pendão, I. Silva, Optical fiber sensors and sensing networks: overview of the main principles and applications. Sensors **22**(19), 7554 (2022). https://doi.org/10.3390/S22197554
22. S.J. Nawaz, S.K. Sharma, S. Wyne, M.N. Patwary, M. Asaduzzaman, Quantum machine learning for 6G communication networks: state-of-the-art and vision for the future. IEEE Access **7**, 46317–46350 (2019). https://doi.org/10.1109/ACCESS.2019.2909490
23. J. Sundaresan, M. Siek, S.M. Tedjojuwono, Sustainable manufacturing through empowering machine tools with IoT innovations: a literature review. IOP Conf. Ser. Earth Environ. Sci. **1324**(1), 012088 (2024). https://doi.org/10.1088/1755-1315/1324/1/012088
24. F. Kruber, J. Wurst, E. S. Morales, S. Chakraborty, M. Botsch, Unsupervised and supervised learning with the random forest algorithm for traffic scenario clustering and classification. IEEE Intell. Veh. Symp., Proc. **2019**(June), 2463–2470 (2019). https://doi.org/10.1109/IVS.2019.8813994
25. Sci-Hub|Reinforcement Learning for Intelligent Healthcare Applications: A Survey. Artificial Intelligence in Medicine, 101964 | https://doi.org/10.1016/j.artmed.2020.101964

Deep Learning Applications in Optical Fiber-Based Sensing Systems

Amit Agrawal and Monu Nath Baitha

Abstract Optical fiber sensor's utilization is gaining momentum in different fields such as telecommunications, medical diagnostics, and structural health monitoring due to their high sensitivity, wide bandwidth, and immunity to electromagnetic interference. The precision and design of these sensors can be further improved with the help of cutting-edge deep learning models which has the capability to significantly amplify the capabilities of the sensors. This chapter delves into the capability of deep learning algorithms, particularly convolutional neural networks (CNNs) and recurrent neural networks (RNNs), long short-term memory (LSTM), and gated recurrent unit (GRU) enhance the performance and functionality of optical fiber sensors. The focus of this chapter is implementation of deep learning for signal processing, pattern recognition, and anomaly detection in fiber optic sensors. The chapter also covers advancements in data fusion techniques, where deep learning combines information from multiple sensors to provide more comprehensive insights compared to conventional techniques. This extensive comparative analysis would lead to more efficient utilization of deep learning techniques for future advancements, ensuring that this technology continues to evolve and meet the growing demands of various industries.

Keywords Optical fiber sensor · Machine learning · Deep learning · RNN · CNN · LSTM · GRU

A. Agrawal
Department of Electrical and Electronics Engineering, ABES Engineering College, Ghaziabad, Uttar Pradesh, India

School of Computer Science & Engineering, IILM University, Greater Noida, Greater Noida, Uttar Pradesh, India

M. N. Baitha (✉)
Department of Information Engineering, University of Brescia, Brescia, Italy
e-mail: monu.baitha@unibs.it

V. Bhardwaj et al. (eds.), *Optical Fiber Sensors and AI*, Progress in Optical Science and Photonics 34, https://doi.org/10.1007/978-981-96-6158-9_6

1 Introduction

The motoring and control of the application and environment in different fields has led to the creation of different sensors in different field of study and analysis [1]. The sensors provide a cost-effective method to monitor and control the processes. The advancement in these sensor technologies depends on various factors and application area. Like sensors are the backbone of automation industry where the robotic systems analyze its environment with the help of sensors and take the decision based on sensor data [2]. The application of sensor is also rapidly increasing in the field of medical and hand-held devises.

The advancement in current sensor technology and storage capacity has paved the way for large data availability for all the spheres of science and technology [3]. This availability of the dataset is crucial for analysis and decision-making, as it provides a window of previous observations for future decision-making [4]. However, the analysis of large amount of dataset is also a challenge due to its complexity and non-linearity. Initially, statistical techniques are used to analyze the useful information from the datasets. But this information was limited to the statistical features of the dataset and fails to provide accurate parameters for more non-linear datasets [5].

Artificial intelligence (AI) techniques are the fastest growing techniques in the current scientific domain [6]. These techniques can improve the performance and accuracy of these optical fiber sensors by improving the classification and prediction accuracy compared to conventional statistical methods. Machine learning and deep learning are the parts of the AI techniques [7]. These techniques rely on the datasets for decision-making.

The vast availability of the data and computational speed has paved the way for machine learning (ML) techniques [8]. ML techniques provide more reliable outcome and prediction compared to statistical techniques. Machine learning models maps the independent feature variable with dependent feature variable to find the exact relationship for the prediction. ML models have gained lot of attention in different fields of study [9]. These models have also been utilized in the field of optical fiber sensor field. In the healthcare system, these methods have utilized to decode the data for effective detection of antibodies. It is also utilized in the field of organic compound classification that improved the result to 92%. Even though ML models have improved the performance of optical fiber sensors, however with the increase in the variable and complexity of the datasets, its performance deteriorates [10]. This limitation comes from its shallow architecture which limits its capacity to generalize the relationship in the latent space dataset.

To address the challenges of ML models, deep learning (DL) models are introduced in the field of optical fiber sensors. The DL models have better capacity in generalizing the input relationship coming from sensor data and will in turn improve the prediction and classification accuracy. The reason for better performance of DL models lies in its architecture. The DL models are deep in nature, which means large number of sequential layers can be attached to a single model.

2 Data Handling

Data is the most crucial aspect of any ML or DL model. It provides a backbone of learning parameters [11]. Similarly handling optical fiber sensor data is a multi-stage optimization that provides a better functional space for accurate and efficient analysis and prediction. The process of data handling starts with the data acquisition from different sources with specific interval [12].

This is critical as it allows the data to be cleaned, and any noises removed while ensuring accurate signal processing. Typically for denoising and extracting essential naturally, methods like Fourier transform is used to solve it [13]. For distributed sensing systems that collect data continuously over the length of an optical fiber, it is crucial to process a huge amount of data in a scalable manner. This has to do with breaking the data into smaller parts and using different sorts of compression algorithms that help them fit on disk.

After preprocessing, the data is mostly processed with machine learning and deep learning techniques. They can search for patterns, abnormal behavior, and predict how the sensors may behave in future which makes this data usable. For example, Convolutional Neural Networks (CNNs) are employed for spatial data analysis from distributed acoustic sensing (DAS) systems, while recurrent neural networks are well suited for time-series data artwork, such as temperature changes detected by distributed temperature wringing (DTS) systems [14].

Importantly, although optical fiber sensor products offer many advantages, data integration and fusion are also necessary to combine the output of such a smart product with that derived from more prevalent sensors for end-to-end monitoring solutions [15]. This is a complete solution which combines all analysis methods and supported alternatives together that improves decision-making power as well provides monitoring and alerting system supporting real-time action. Efficient data management enhances the accuracy and stability of optical fiber sensors as well as extends their capacity, enabling valuable information in a wide scope of fields (Fig. 1).

3 Algorithm Comparison

3.1 Training Techniques

This sector of the work describes the training methods of the DL model. This can be broadly categorized in three sub-sections: supervised, semi-supervised, and unsupervised.

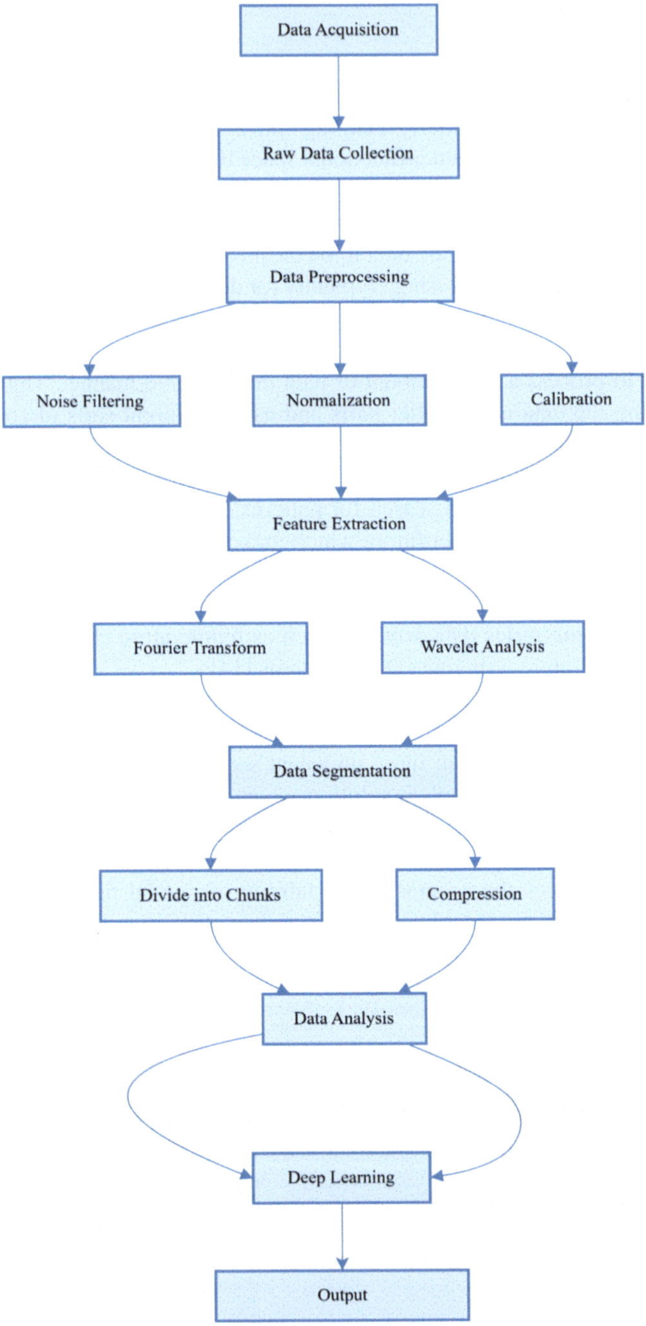

Fig. 1 Flowchart of deep learning models for optical fiber-based data

3.2 DL Model Types

The DL models are based on neural network [16]. Every model has different type of arrangement of neural network. Initial research works have focused on artificial neural network (ANN) and deep neural network (DNN). These DL models have better performance compared to ML models due to their deep architecture. During training process every DL model adjusts the weight and bias of the model [4]. These weights and biases of every connection of different layers form the functional space of learning parameters.

A. **Recurrent Neural Network**

Feedforeword Recurrent Neural Network (RNN) is another type of neural networks designated for handling data that arrive in a sequence, thus they fit to the characteristics adopted by optical fiber sensing systems—such systems usually provide time-series-based information. RNNs have a hidden state that propagates information from previous time steps and enables RNN to model temporal dependencies, as shown in Fig. 2. This is especially handy for finding patterns and anomalies in sensor data through time [17]. So, RNNs are a good fit as they learn over time, but there are problems with them—e.g., vanishing gradients—that make it hard to model long-term dependencies in the many sequences of optical fiber sensor data.

B. **Long Short-Term Memory (LSTM)**

LSTMS are a special kind of RNN, capable to learn long-term dependencies and avoid short memory issues (like gradient vanishing). They also have a memory cell and three gates—input, forget, and output gates that control the flow of information which make them better at catching long-term dependencies [18]. This attribute makes LSTMs extremely helpful when analyzing optical fiber sensor signal data because long-range trends and subsequence correlations that are essential for reliable monitoring and detection of anomalies can be captured well, as shown in Fig. 3. This feature is particularly useful for applications such as Structural Health Monitoring (SHM) using optical fiber sensors.

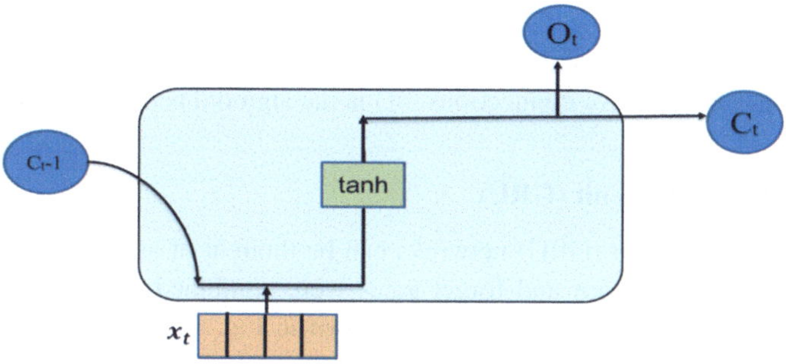

Fig. 2 RNN architecture

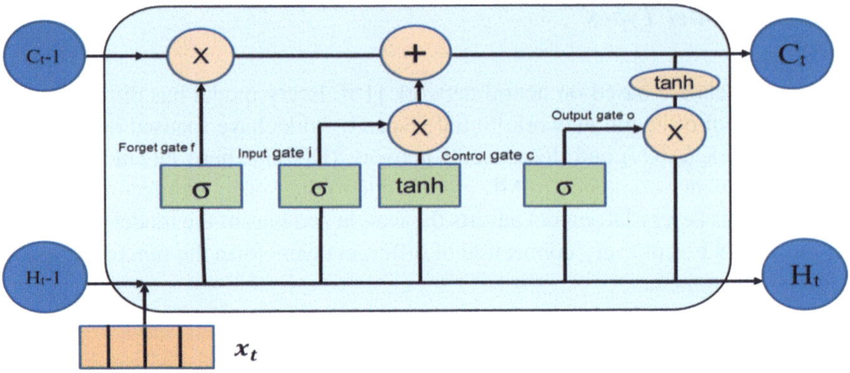

Fig. 3 LSTM architecture

CNN uses a memory cell unit $GM^{\langle t\rangle}$ that stores the sequential information of current state with latent information of previous state. Forget gate decides the relevance of the previous information state while taking the decision for next state. The state is depicted as

$$GM^{\langle t\rangle} = \sigma\left(W_f\left[GM^{\langle t-1\rangle}, X^{\langle t\rangle} + B_c\right]\right) \tag{1}$$

$$P^{\langle t\rangle} = \sigma\left(W_i\left[GM^{\langle t-1\rangle}, X^{\langle t\rangle} + B_i\right]\right) \tag{2}$$

$$C'^{\langle t\rangle} = tanh\left(W_c\left[GM^{\langle t-1\rangle}, X^{\langle t\rangle} + B_c\right]\right) \tag{3}$$

$$C^{\langle t\rangle} = GM^{\langle t-1\rangle} * C^{\langle t-1\rangle} + P^{\langle t\rangle} * C'^{\langle t\rangle} \tag{4}$$

$$A^{\langle t\rangle} = \sigma\left(W_\circ\left[GM^{\langle t-1\rangle}, X^{\langle t\rangle} + B_\circ\right]\right) \tag{5}$$

$$h^{\langle t\rangle} = A^{\langle t\rangle} * tanh\left(C^{\langle t\rangle}\right) \tag{6}$$

where W_f, W_c, W_o, and W_i are the weights of the connections and B_c, B_o, B_i, and B_t are the biases of respective connections. σ() is the sigmoid activation function with the link.

C. **Gated Recurrent Unit (GRU)**

Gated Recurrent Unit (GRU) networks can be thought of as simplified versions of LSTMs where the input and forget gates were combined into a single update gate and output gating was removed, As shown in Fig. 4, we can build a lighter architecture with much smaller number of parameters and still keep the ability to store information about longer term dependencies in sequential data [19]. The simpler

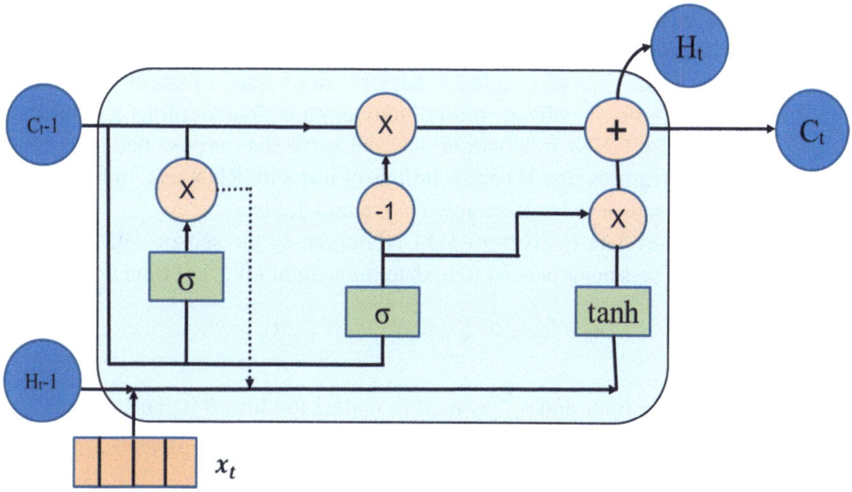

Fig. 4 GRU architecture

structure and reduced computational requirements make GRUs better equipped in processing optical fiber sensor data. Also, they are performing almost equal efficiency which makes them compatible with quick and precise analysis of sensor data for real-time monitoring applications.

The input data is given to GRU layer. The present state is represented as

$$h^{\langle t\rangle} = A^{\langle t\rangle} * \tanh C^{\langle t\rangle} \tag{7}$$

$$R_v = \sigma\left(W_r\left[GM^{\langle t-1\rangle}, X^{\langle t\rangle} + B_r\right]\right) \tag{8}$$

$$G^{\langle t\rangle} = U_g * GM^{\langle t\rangle} + \left(1 - U_g\right) * G^{\langle t-1\rangle} \tag{9}$$

$$U_g = \sigma\left(W_u\left[G^{\langle t-1\rangle}, X^{\langle t\rangle} + B_u\right]\right) \tag{10}$$

$$A^{\langle t\rangle} = g\left(G^{\langle t\rangle}\right) \tag{11}$$

where R_v is a relevance gate which gives a weight to $GM^{\langle t-1\rangle}$, which in turn decides its role in determining the current state with tanh non-linear activation function with bias B_r and weight W_r. And update gate is U_g that goes to next gate for output calculation.

D. **Convolution Neural Network (CNN)**

Although Convolutional Neural Networks (CNN) are widely used in spatial data, they have also been applied to time-series data analysis, for instance, optical fiber sensor data. CNNs use convolutional filters to highlight features in the data, and local patterns are essential for detecting events or anomalies [20]. In the context of optical fiber sensor data, CNNs are well suited to detect various spatial patterns within this high-dimensional input—localized strain or temperature changes along a thousand-meter-long fiber. But, CNNs in isolation may not serve the purpose better because of needing special features like 1D convolutions or use with RNN, etc. to handle the temporal nature of sensor data effectively (Fig. 5 and Table 1).

The pre-processed data is given to 1D CNN layer, as the sensor data lie in the sequential frame. The sensor is used to update the weight ($W^{\langle t \rangle}$) at time t:

$$W^{\langle t \rangle} = F_i^{\langle L \rangle} * F_i^{\langle L \rangle} * n_c^{\langle L-1 \rangle} * n_c^{\langle L \rangle} \tag{12}$$

where $F_i^{\langle L \rangle}$ shows the filter and $n_c^{\langle L \rangle}$ is used to update the bias $B^{\langle t \rangle}$. This weight and bias leads to output of CNN as

$$C^{\langle t \rangle} = W^{\langle t \rangle} * A^{\langle t-1 \rangle} + B^{\langle t \rangle} \tag{13}$$

RNNs, LSTMs, GRUs, and CNNs each offer unique strengths for analyzing optical fiber sensor data. RNNs are suited for capturing temporal dependencies but struggle with long sequences due to vanishing gradients. LSTMs address this with a

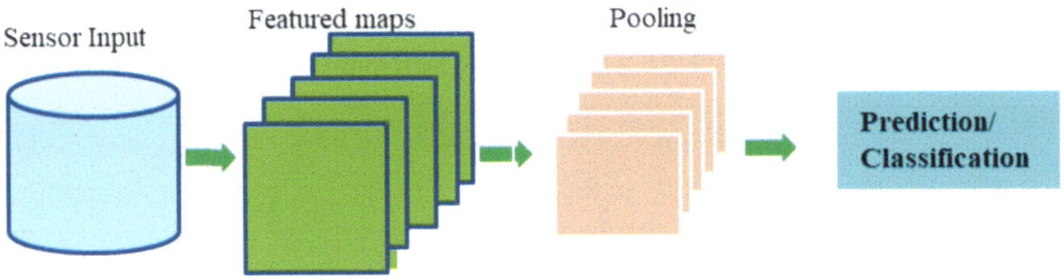

Fig. 5 CNN architecture

Table 1 Comparison of DL model

Model type	Number of parameters	Processing time	Gates	Key use case
CNNs	High (up to hundreds of millions)	High for training	N/A	Spatial data analysis (e.g., vibrational events)
RNNs	Moderate	Slow due to sequential nature	Simple activations	Sequential data (e.g., time series)
LSTMs	High	Moderate	Input, forget, output	Long-term dependencies in time series
GRUs	Moderate	Faster than LSTMs	Reset, update	Time series with simpler architecture

more complex architecture that effectively captures long-term dependencies, making them ideal for detailed monitoring tasks. GRUs offer a more efficient alternative to LSTMs, providing a balance between complexity and performance. CNNs excel at identifying spatial patterns within the sensor data but may need to be combined with recurrent networks to handle temporal aspects comprehensively. For applications requiring real-time analysis and long-term dependency capture, LSTMs or GRUs are typically preferred, while CNNs are beneficial for tasks emphasizing spatial feature extraction.

4 Scalability of DL Models on Optical Fiber Sensor Systems

Deep Learning (DL) models must be able to scale up on optical fiber sensor systems toward efficient data processing and practical application in different domains. The block diagram below shows how this pipeline connects sensor environments to a variety of applications including healthcare, power plants, telecommunications, and medical diagnostics.

Implementation of DL Models

Data Collection and Preprocessing: It starts with collecting sensor data that itself is massive in size and complexity. Preprocessing these data is important, to cover noise removal, normalization, and dimensionality reduction. This stage results in having the data clean and standardized to be fitted into the DL models.

Function Mapping and Generalized Function: Postprocessed data is given as input to DL models in order to perform function mapping where the model learns about its pattern's recognition, anomalies, etc. Such mapping leads to a kind of universal function, which can be further used across different use cases (Fig. 6).

Scalability and Performance of the Model

High accuracy: All models achieve high quantitative performance, a necessity for accurate detection both in situations and diagnostics. This ensures that the data models can accurately interpret sensor input and yield actionable insights.

Speed: As machine comes from the different models, CNNs and LSTMs are in between but its GRUs that outperform them all time-wise. This is crucial for needs like live monitoring or time-critical applications (e.g., medical diagnosis, power plants).

We narrowed them down to four categories: Scalability—as these models must be deployed at a very large scale. However, CNNs and GRUs have much more scalability potential with respect to RNNs and LSTMs. With their reduced architecture, GRUs strike a balance between training speed and large-scale time-series data analysis in the context of telecommunication networks.

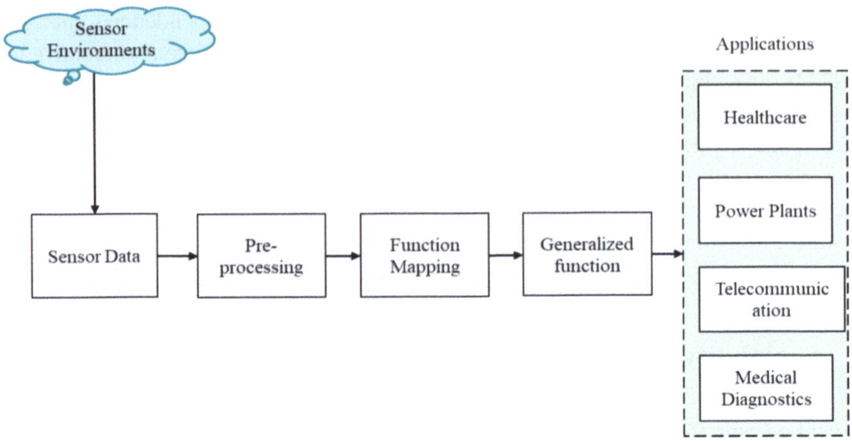

Fig. 6 Structure of sensor environment

Table 2 Comparison of deep learning models

Model type	Key use case	Data type	Accuracy	Speed	Scalability	Implementation complexity
CNNs	Spatial event detection	Spatial/ temporal	High	Moderate	Good	High
RNNs	Sequential data analysis	Time series	High	Slow	Moderate	Moderate
LSTMs	Long-term temporal dependencies	Time series	High	Moderate	Moderate	High
GRUs	Simplified RNN for time series	Time series	High	Faster	Better	Moderate

Details Flow: The complexity of the details flow affects both deployment and maintenance. The higher implementation complexity of CNNs and LSTMs, e.g., because they are more sophisticated. GRUs and RNNs, on the other hand, have moderate complexities which allows them to be deployed and managed across multiple sensor environments (Table 2).

5 Conclusion

The use of Deep Learning (DL) models in fiber optic sensing systems is a major step forward for sensor technology. In this chapter, we discuss how the DL architectures, especially CNNs, RNNs LSTMs, and GRUs, can take care of complex feature engineering in multichannel or voluminous sensor data concerning a different range of applications. CNI models have very high precision in signal processing, pattern recognition, and anomaly detection—skills that are crucial for use cases across healthcare, power plants, telecommunications, and medical diagnostics.

CNNs are best for detecting spatial patterns making them ideal in problems which involves analyzing of spatial patterns. Even with their lower processing speed compared to other model types, RNNs remain relevant when dealing with time-series data due to the way they process sequences. While they are more difficult to implement as compared to the CNN, LSTMs have been shown to better capture long-range dependencies in data which is important for fine-grained risk monitoring. Because it provides a desirable trade-off between efficiency and performance, GRU is mundane when directly used in real-time applications that require faster processing speed with large-scale scalability.

There is a systematic pipeline of data collection and preprocessing, function mapping to generate generalized functions that serve over diverse use cases integrates these models into optical fiber sensor system. It means that these models are scalable and can easily support large-scale deployments giving high-quality insights which will be valuable to make decisions.

Finally, the ultimate horizon will empower optical fiber sensors with DL models to enhance their performance and capability thereby creating a model for accurate monitoring systems in different sectors. This evolution of these models continuously provides newer advancements to many aspects in the field of sensor technology which aim future-oriented anticipation that meets increased demands for different markets.

Acknowledgements We are deeply grateful to Dr. Amit Rai for his invaluable guidance and support throughout the preparation of this book chapter. His expertise and mentorship were instrumental in shaping our research and ensuring the quality of our work.

References

1. U.M.N. Jayawickrema, H.M.C.M. Herath, N.K. Hettiarachchi, H.P. Sooriyaarachchi, J.A. Epaarachchi, Fibre-optic sensor and deep learning-based structural health monitoring systems for civil structures: a review. Measurement **199**, 111543 (2022)
2. P. Roriz, O. Frazão, A.B. Lobo-Ribeiro, J.L. Santos, J.A. Simões, Review of fiber-optic pressure sensors for biomedical and biomechanical applications. J. Biomed. Opt. **18**, 050903 (2013)
3. B.J. Shastri, A.N. Tait, T. Ferreira de Lima, W.H.P. Pernice, H. Bhaskaran, C.D. Wright, P.R. Prucnal, Photonics for artificial intelligence and neuromorphic computing. Nat. Photonics **15**, 102 (2021)
4. W. Ma, Z. Liu, Z.A. Kudyshev, A. Boltasseva, W. Cai, Y. Liu, Deep learning for the design of photonic structures. Nat. Photonics **15**, 77 (2021)
5. H.-N. Li, D.-S. Li, G.-B. Song, Recent applications of fiber optic sensors to health monitoring in civil engineering. Eng. Struct. **26**, 1647 (2004)
6. J. Varanasi, M.M. Tripathi, Artificial neural network based wind speed & power forecasting in US wind energy farms, in *2016 IEEE 1st International Conference on Power Electronics, Intelligent Control and Energy Systems (ICPEICES)* (IEEE, 2016), pp. 1–6

7. C.R.U. Kumari, D. Samiappan, K.R., T. Sudhakar, Fiber optic sensors in ocean observation: a comprehensive review, Optik (Stuttg) **179**, 351 (2019)
8. M.M. Malik, A hierarchy of limitations in machine learning, ArXiv **abs/2002.0**, (2020)
9. H.F. Pardede, E. Suryawati, D. Krisnandi, R.S. Yuwana, V. Zilvan, Machine learning based plant diseases detection: a review, in *2020 International Conference on Radar, Antenna, Microwave, Electronics, and Telecommunications (ICRAMET)* (IEEE, 2020), pp. 212–217
10. V.A. Zhmud, N.O. Kondratiev, K.A. Kuznetsov, V.G. Trubin, L.V. Dimitrov, Application of ultrasonic sensor for measuring distances in robotics. J. Phys. Conf. Ser. **1015**, 032189 (2018)
11. Y. LeCun, Y. Bengio, G. Hinton, Deep learning. Nature **521**, 436 (2015)
12. R.D.Q.D. Lqglu, D. Lv, D. Frqvlghuhg, I.R.U. Wkh, D. Dqdo, S. Lq, W.K.H. Olwhudwxuh, V.D.U. Pdq, G. Lq, A review of data mining and solar power prediction **0**(3) (2016)
13. P. Ray, R.K. Lenka, M. Biswal, Frequency mode identification using modified masking signal-based empirical mode decomposition. IET Gener. Transm. Distrib. **13**, 1266 (2019)
14. B. Zhong, X. Xing, P. Love, X. Wang, H. Luo, Convolutional neural network: deep learning-based classification of building quality problems. Adv. Eng. Inform. **40**, 46 (2019)
15. U. Utzinger, R.R. Richards-Kortum, Fiber optic probes for biomedical optical spectroscopy. J. Biomed. Opt. **8**, 121 (2003)
16. H. Shi, M. Xu, R. Li, Deep learning for household load forecasting—a novel pooling deep RNN. IEEE Trans. Smart Grid. **9**, 5271 (2018)
17. H. Shi, M. Xu, R. Li, Deep learning for household load forecasting-a novel pooling deep RNN. IEEE Trans. Smart Grid. **9**, 5271 (2018)
18. S. Hochreiter, J. Schmidhuber, Long short-term memory. Neural Comput. **9**, 1735 (1997)
19. J. Chung, C. Gulcehre, K. Cho, Y. Bengio, Empirical evaluation of gated recurrent neural networks on sequence modeling (2014), arXiv:1412.3555
20. S. Song, H. Huang, T. Ruan, Abstractive text summarization using LSTM-CNN based deep learning. Multimed Tools Appl. **78**, 857 (2019)

AI-Driven Design and Optimization of Optical Fiber Sensor Networks

Garima Chanana

Abstract In recent years, the convergence of artificial intelligence and optical fiber sensor networks has revolutionized sensor technology, significantly enhancing performance, reliability, and efficiency. Optical fiber sensor networks, known for their high sensitivity, immunity to electromagnetic interference, and extensive bandwidth capacity, facilitate precise and distributed sensing of various physical parameters. These networks are pivotal in industries such as oil and gas, aerospace, environmental monitoring, biomedical diagnostics, and structural health monitoring. AI algorithms are useful in advancing these networks and providing capabilities beyond traditional systems. This study explores AI-driven methodologies that can augment the capabilities of optical fiber sensor networks across various domains. By transforming sensor data into actionable insights, AI can foster advancements in predictive analytics, operational efficiency, and decision-making. Embracing AI methodologies unlocks new potentials for innovation, sustainability, and reliability in sensor network applications, fostering smarter and more resilient technological solutions. Integrating these two domains will result in a transformative leap, enabling sophisticated monitoring, enhanced safety, and efficient management across industrial and environmental domains. This chapter encompasses the synergy of AI and sensor technology that will be useful in driving significant progress, paving the way for smarter, more sustainable solutions in the modern era.

Keywords AI-driven sensing · Optical fiber sensor networks · Predictive analytics · Real-time monitoring

G. Chanana (✉)
School of Engineering and Technology, Vivekananda Institute of Professional Studies-Technical Campus, Delhi, India
e-mail: garima.chanana@vips.edu

V. Bhardwaj et al. (eds.), *Optical Fiber Sensors and AI*, Progress in Optical Science and Photonics 34, https://doi.org/10.1007/978-981-96-6158-9_7

1 Introduction to Optical Fiber Sensor Networks

Optical Fiber Sensor Networks (OFSNs) [1] are a sensing system that uses optical fibers as the main medium for detecting and transmitting information. These networks integrate various sensors within the optical fibers or utilize the fibers themselves as sensing elements to capture changes in various parameters such as temperature, pressure, strain, or chemical compositions. OFSNs are characterized by their ability to provide distributed sensing, allowing for the monitoring of multiple points along the fiber's length. This is particularly useful for large-scale and continuous monitoring applications. The high sensitivity and accuracy of optical fibers enable precise measurements, while their immunity to electromagnetic interference makes them ideal for use in noisy environments. Additionally, optical fibers can transmit information over long distances with minimal signal loss, which is useful for monitoring extensive areas. Their durability and reliability ensure consistent performance even under extreme conditions. As a result, OFSNs are widely employed in various fields ranging from oil and gas [2], spacecraft applications [3], environmental sensing [4], biomedical diagnostics [5], structural health monitoring [6], and telecommunications [7].

In recent years, the synergistic convergence of artificial intelligence (AI) and OFSNs has brought a paradigm shift in how we perceive and utilize sensor technology. In this era, by leveraging the sophistication of AI-driven algorithms, it is possible to enhance the performance, reliability, and efficiency of OFSNs for diverse sets of applications. Incorporation of AI algorithms with OFSNs empowers these networks with capabilities beyond traditional sensor systems. AI-driven OFSNs allow for predictive maintenance by analyzing historical data, which is useful to forecast equipment failures [8], enhancing operational reliability. Moreover, anomaly detection algorithms [9] continuously monitor sensor data of OFSNs to detect deviations from normal operating conditions, ensuring early identification of faults and timely maintenance actions.

When constructing OFSNs, it is essential to consider multiple aspects to ensure their performance and reliability. One of the primary factors is the selection of optical fibers. Typically, single-mode fibers are chosen for their superior sensitivity and ability to cover long distances, while multimode fibers are suitable for shorter distances but higher power applications [10]. Additionally, the physical characteristics of the fibers, such as their resilience to environmental conditions like temperature and pressure, are important. Fiber selection should be based on its robustness in the target environment, such as for biomedical [11] or marine applications [12]. Furthermore, the network topology and architecture also play a significant role [13]. It has also been observed that employing multiplexing techniques [14–16] can greatly boost the network's capacity and efficiency. These methods enable the concurrent transmission of multiple signals through a single fiber, enhancing the network's data handling capability and supporting numerous sensors.

When developing OFSNs, it's essential to integrate advanced signal processing techniques [17] and interrogation methods [18]. This design must employ sophisticated algorithms and high-resolution detection systems to ensure the optical signals from the sensors are interpreted accurately. Such precision is crucial for obtaining reliable data, which is vital for practical applications. Additionally, it's important to factor in costs, installation, and maintenance. The initial setup may be costly and intricate, and might also necessitate specialized equipment and skills. However, the long-term advantages, such as low maintenance expenses and high durability, typically justify the investment.

In summary, designing effective OFSNs involves careful selection of fiber types, employing advanced network topologies and signal processing techniques, and considering the integration with other technologies to enhance performance and functionality. These factors collectively ensure that the network can meet the demanding requirements for various applications.

In this chapter, the objective is to review the AI-based design and optimization strategies that can be integrated into OFSNs to boost their performance. By utilizing AI technologies, various aspects of OFSNs can be optimized. The incorporation of AI can result in smarter, more efficient, and dependable sensor networks that are capable of fulfilling the diverse and stringent requirements of contemporary applications. Section 2 introduces the operating principles of optical fiber sensing networks, providing a foundation for understanding how these networks work and their various types and applications. Section 3 presents a brief discussion on Machine Learning (ML) and Deep Learning (DL), highlighting their key concepts, methodologies, and relevance to modern data processing and analysis suitable for OFSNs. Section 4 will delve into state-of-the-art AI-integrated OFSNs, offering insights into how AI can drive innovations and advancements in the design and functionality of these sophisticated sensing systems. Section 5 concludes with a discussion and future perspectives, reflecting on the current state of research, potential challenges, and opportunities for further advancements in integrating deep learning with optical sensor technology.

2 Operating Principles of Optical Fiber Sensing Networks

Optical fibers are fine strands composed of typically either glass or plastic material, which are designed to carry light signals over extended distances with minimal signal degradation. They are primarily classified into two types: single-mode and multimode fibers. Single-mode fibers allow only one light mode to propagate, making them ideal for long-distance communication with higher bandwidth. Multimode fibers support multiple light modes, making them suitable for short-distance applications [19].

Using these optical fibers, OFSNs can be developed that use optical fibers as the medium for transmitting light signals. These are used to monitor and measure various physical, chemical, and environmental parameters. These networks leverage optical sensors that are embedded along or at the ends of the fiber to detect changes in the environment by measuring variations in the light's properties, such as intensity, phase, wavelength, or polarization.

These networks usually begin with a light source, typically a laser, generating an optical signal that travels through the optical fiber. This light reaches optical sensors, which interact with the light based on environmental changes like strain or temperature, causing a wavelength shift. The modulated light then returns through the fiber to an optical signal processing unit, where it is converted into an electrical signal, demodulated, and processed to extract relevant information. In some cases, multiplexing allows multiple sensors to be monitored along the same fiber. The processed signal is then fed into a data acquisition system, which collects and converts it into digital data. This data undergoes signal analysis and calibration, correlating wavelength shifts with physical parameters. Finally, the processed data is visualized in real time using monitoring tools, providing actionable insights for further decision-making (Fig. 1).

There exist various optical fiber sensors [20] that utilize different principles and structures to detect changes in environmental parameters. Fiber Bragg Gratings (FBGs) [21] are created by introducing periodic variations in the refractive index within the core of the optical fiber. This grating reflects specific wavelengths of light while allowing others to pass through. The reflected wavelength is known as Bragg's wavelength, which changes in response to external physical perturbations such as strain, temperature, and pressure, making FBGs highly useful for sensing

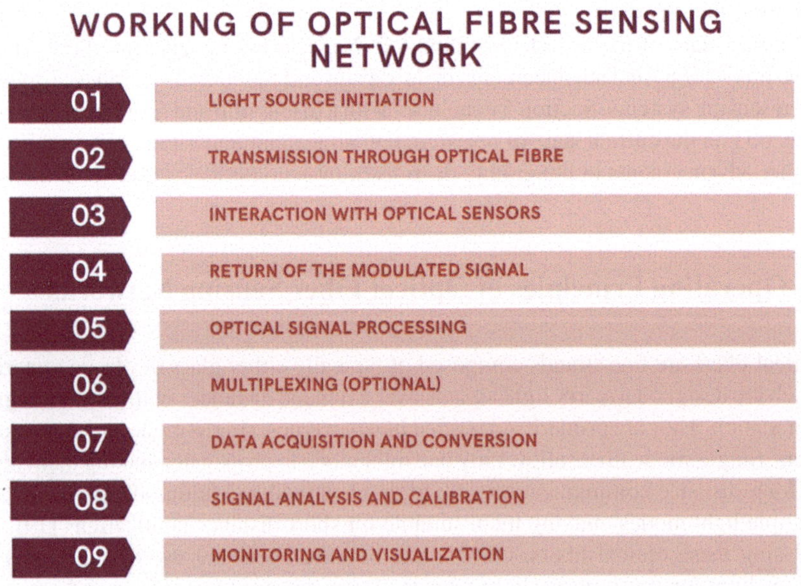

Fig. 1 Working of optical fiber sensing network

applications [22]. Long-Period Fiber Gratings (LPFGs) [23] are defined by periodic refractive index variations along the fiber core, with grating periods that are much longer than those in FBGs. In LPFGs, the typical grating period is on the order of hundreds of micrometers, as opposed to a few micrometers in FBGs. This long period enables coupling between the modes of core and cladding. These are sensitive to variations in refractive index and therefore can be used for biological and chemical sensing [24]. Another type of optical fiber is the Tapered Optical Fiber Interferometer (TOFI), which uses a tapered section of the fiber to create interference patterns, providing high sensitivity for refractive index and biochemical protein sensing [25].

Fabry–Pérot Interferometer Fiber (FPIF) sensor [26] is based on the phenomenon of multiple light reflections within a cavity, leading to interference, and is used in pressure sensing [27] and displacement sensing [28]. Another interferometer fiber sensor called the Fiber-Optic Mach–Zehnder Interferometer (FOMZI) [29] splits light into two arms and measures the phase difference, which is used in strain, temperature, and acoustic sensing. Fiber-Optic Michelson Interferometer (FOMI) [30] functions similarly to Mach–Zehnder but with a single arm acting as a reference. It is used in precision measurements of vibrations and displacements [31].

Another type is the Sagnac Interferometer [28], which exploits the Sagnac effect, i.e., where light traveling in opposite directions around a loop interferes. It is suitable for rotation and angular velocity measurements [32]. Surface Plasmon Resonance (SPR) based sensors use the interaction of light with metal-dielectric interfaces to detect minute changes in refractive index. It is ideal for environmental sensing [33]. Fluorescence-based Fiber-Optic Probes (FFOP) use the emission of fluorescent light in response to a target and are used as biosensors [34]. The Light Diffuser-based Optical Fiber Sensors (LDOF) incorporate light diffusers into the fiber to enhance light interaction with the surrounding medium. They are useful in medical applications [35] (Fig 2).

Comparing these sensor approaches reveals that interferometer-based sensors provide high precision and sensitivity, while FBGs and LPFGs are compact and well-suited for distributed sensing. SPR and fluorescence-based probes are highly sensitive to biochemical changes, making them ideal for medical and chemical applications. Each sensor type is chosen based on the specific requirements of the application, such as sensitivity, range, and environmental conditions.

An optical fiber sensing network can be developed by integrating different optical fiber sensors that are capable of measuring different environmental and structural parameters in real time. OFSNs [36] using these sensors operate based on the principle of light transmission through optical fibers. They utilize light-environment interaction to detect changes in the environment, such as temperature, strain, and vibration. OFSNs integrate different types of optical fiber as discussed above, which can be positioned at multiple locations along the fiber, enabling distributed sensing that can cover long distances with minimal signal loss.

The OFSNs offer several advantages, including immunity to electromagnetic interference, high sensitivity, long-distance data transmission, and the ability to function in harsh environments [1]. By integrating multiple sensors along a single fiber, these networks provide a cost-effective, scalable, and highly reliable solution for

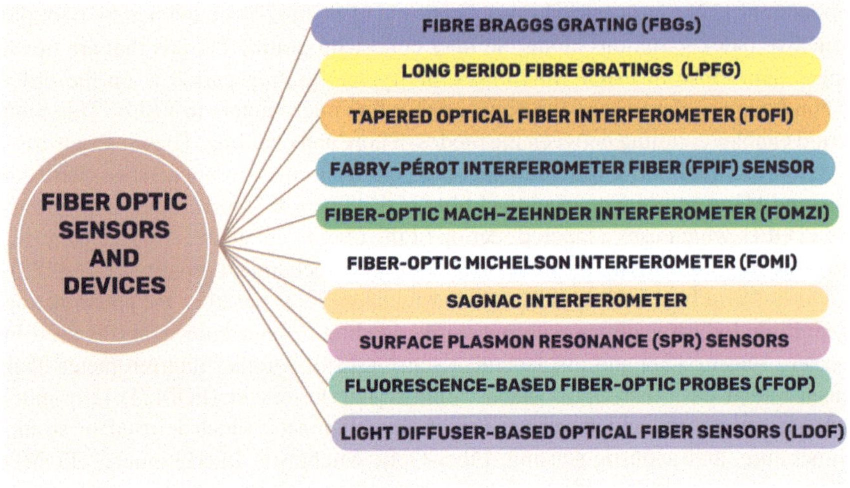

Fig. 2 Types of fiber-optic sensors and devices

real time, distributed sensing. With advancements in technology, OFSNs continue to expand in capability, providing a robust, efficient, and scalable solution for modern sensing needs.

3 Artificial Intelligence (AI) Techniques for OFSN

This section explores the application of various machine learning and deep learning techniques to OFSNs and highlights how these methods improve performance, efficiency, and management. Machine Learning (ML) [37] plays a key role in enhancing the capabilities of OFSNs by processing large amounts of sensor data, improving accuracy, and automating the interpretation of signals from sensors like FBGs, LPFGs, and different interferometer-based optical fibers. In these networks, ML helps analyze and predict sensor responses to environmental factors such as strain and temperature [38].

Supervised learning algorithms, such as support vector machines (SVM), have been successfully applied for data detection in coherent optical fiber 16-APSK communication systems, effectively addressing nonlinear phase noise, outperforming traditional maximum-likelihood detection [39]. A study by Yigit et al. [40] introduced support vector regression (SVR) to analyze the diaphragm of a custom Fabry–Perot sensor. By training the SVR model on 1,512 designs, they successfully predicted dynamic and static pressure responses in real time. ML algorithms have also been effective in detecting and locating faults in OFSNs [41]. Models like SVM, CNN, LSTM, and decision trees achieved accuracy rates between 86% and 98%, outperforming traditional methods [42].

In another study, Mahani et al. [43] developed an automated design framework using Finite-Difference Time-Domain (FDTD) simulations for rectangular surface Bragg gratings in semiconductor diode lasers. They applied ML to predict reflectance spectra, and the optimized XGBoost model achieved over 99% accuracy, greatly reducing computational time compared to FDTD simulations.

Another phenomenon called the nonlinear phase noise (NLPN) is also studied in optical transmission systems. This process is caused by signal-noise interaction during the transmission of data and has been widely studied. Study by Wang et al. [44] demonstrated that SVM is highly effective in mitigating this NLPN. It outperformed maximum-likelihood estimation (MLE) by reducing the bit-error rate (BER), expanding the launch power dynamic range, and extending transmission distance.

Deep Learning (DL) methods [45] offer an advanced approach to OFSNs by learning complex, nonlinear relationships between sensor signals and external conditions. DL techniques are particularly effective in processing high-dimensional sensor data. CNNs are frequently used to detect patterns in spectral responses from sensors. RNNs and their variants are particularly effective for analyzing time-series data, tracking sensor output changes over time, and identifying long-term trends.

Research by Tan et al. [46] combined a Mach–Zehnder optical fiber sensor with CNNs to enhance curvature sensing capabilities. This combination extended the sensor's curvature measurement range to 0–10.4167 m^{-1} and allowed accurate measurements by analyzing the transmission spectrum. This advancement significantly improves the versatility of OFSNs for curvature sensing. Wei et al. [47] introduced a method that integrates CNNs with LSTM networks to demodulate fiber-optic Fabry–Perot (FP) strain sensor spectra. This approach eliminates the need for traditional peak tracking algorithms by directly extracting strain information from normalized spectra. It achieves an impressive accuracy ($R^2 > 99.99\%$) and a low RMSE of 0.524 $\mu\varepsilon$, demonstrating its effectiveness in reducing costs and enhancing the performance of sensing systems.

In another study, Mei et al. [48] utilized a CNN model to address spectral overlap issues in a pressure sensor using a Sagnac interferometer sensor with side-hole fiber (SHF). The model demonstrated high accuracy ($R^2 = 99.99987\%$) and a low RMSE. This highlights its potential for cost-effective sensing solutions even with low-resolution devices. Yamaguchi et al. [49] proposed a CNN-based method for detecting peaks in FBG sensors to improve wavelength-division multiplexing. Their approach enhanced peak wavelength detection accuracy from 30.2 pm to 2.8 pm, making it possible to measure overlapping FBGs more efficiently and economically.

Liu et al. [50] developed a method for detecting central wavelengths in serial wavelength-division multiplexing FBGs networks using superimposed FBGs to train a CNN. This technique effectively analyzes overlapping spectra, achieving an RMSE of 1.819 pm and a demodulation time of under 53.86 ms. Building on the advancements in optical fiber sensing, Nyugen et al. [51] proposed a deep neural network approach to process complex optical interference from multimode optical fibers even in noisy environments. This method bypasses the need for physical shielding or coatings, providing accurate measurements under challenging conditions. Augmenting these innovations, Ali et al. [52] introduced a dual-stage deep learning model for

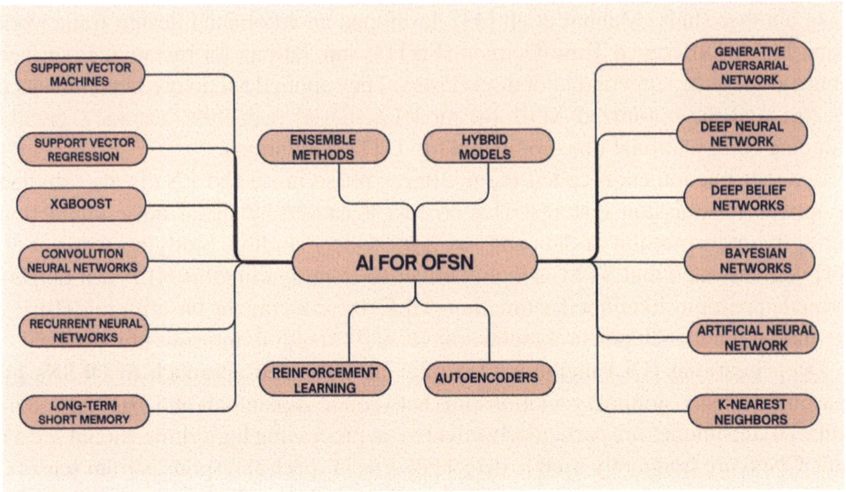

Fig. 3 Artificial intelligence techniques for optical fiber sensing networks (OFSN)

detecting and localizing multiple events in a Sagnac interferometer optical fiber system. This model combines multi-class classification for event detection with deep learning regression for precise localization, demonstrating high accuracy across various signal-to-noise ratios and improving the overall effectiveness of optical fiber sensing systems.

Li et al. [53] developed a cost-effective demodulation system by combining LPFG with artificial intelligence. Their approach integrates a GAN for data augmentation and a DNN for precise wavelength interrogation. They achieved a wavelength precision of ± 3 pm. This innovative solution provides high performance and affordability for fiber-optic sensor applications, with a broad demodulation range Fig. 3 illustrates the various AI techniques that can be employed in AI-based optical fiber sensing networks.

Building on advancements in sensor technology, Wang et al. [54] introduced a novel temperature demodulation method that leverages Deep Belief Networks (DBN) and ensemble learning. Addressing the challenges of irregular changes in thermo-optical and thermal expansion coefficients, their technique uses a stacking ensemble of five DBN models with diverse activation functions. This method attained better demodulation precision over a temperature range of 30 to 1100 °C, thereby enhancing the precision and reliability of fiber-optic sensing systems in high-temperature environments.

In a related advancement, Karanov et al. [55] demonstrated how end-to-end deep learning can enhance optical communication systems. Their approach achieved a data rate of 42 Gb/s over 40 km, maintaining error rates below the required threshold. This performance surpasses traditional methods, highlighting the potential of deep learning to drive significant improvements in optical communication technologies.

This section outlines how different machine learning and deep learning enhance optical fiber sensing networks by improving accuracy, efficiency, and fault detection. ML and DL techniques offer advanced solutions for signal interpretation and real-time analysis, driving significant improvements in performance and cost-effectiveness.

4 Harnessing Artificial Intelligence for Enhanced Optical Fiber Sensing Networks

The integration of AI with OFSNs is transforming the landscape of sensor technology, enhancing both the functionality and efficiency of these systems. AI technologies, particularly ML and DL [56], have been increasingly applied to analyze the vast amount of data generated by OFSNs, driving innovations in sensor design and data processing. One of the most significant contributions of AI to OFSN is in the area of data processing and interpretation [57]. Optical fiber sensors generate large datasets that often contain noise and require complex processing to extract meaningful information, which can be used for failure management [58]. Traditional data analysis techniques can be limited in handling these challenges, especially in real-time applications. AI algorithms, particularly machine learning, can be trained to recognize patterns and anomalies in sensor data [59], improving the accuracy and reliability of measurements.

For instance, ML techniques of SVM have been successfully applied to identify and classify signals from FBGs [60]. A study by Raju et al. [61] Tilted Fiber Bragg Grating (TFBG) sensor, coated with an optimized material, was developed for simultaneous humidity and temperature measurement. The ML method of gradient boosting has been trained on spectrum data and is shown to achieve results of absolute error of 0.6509, RMS of 2.7412, and MAPE of 1.312.

AI has also enabled advanced signal processing techniques in OFSNs, allowing for more sophisticated analysis of signals [62]. LSTMs [63] have been utilized for time-series analysis in optical fiber sensing. A significant example is the study by Hou et al. [64], which introduces a tapered optical fiber-based distributed sensor capable of monitoring contact position and force. By applying a multi-scale CNN+LSTM to the output signal, they achieved position and force prediction resolutions of 7.6 μm and 0.02 N, respectively, over a 5.6 mm length.

AI can also contribute to the intelligent design and optimization of OFSNs. By leveraging AI algorithms, researchers can optimize sensor placement [65], configure network topology, and improve sensor fabrication processes to enhance sensitivity and reduce costs. One such study was performed by Dogan et al. [66] that optimize the parameters of a fiber-optic SPR refractive index sensor using FEM simulations and ANN model. By optimizing an Ag-grated D-shaped sensor, the highest sensitivity of 3890 nm/RIU was achieved using a Genetic Algorithm, with input parameters of a 10 nm air gap, 20 gratings, 50 nm residual cladding, and 75 nm silver layer thickness.

The model good R^2 score of 0.98 for the training dataset. In another study by Kaur et al. [67], the effect of strain on civil structures is investigated using FBG sensors with the help of a genetic algorithm, which optimizes key parameters, including Poisson's ratio and photoelastic coefficients (P11 and P12), to enhance sensor sensitivity.

Another area where AI is making significant strides is in predictive maintenance and fault detection. AI-driven predictive models can analyze data from optical fiber sensors to forecast potential failures or maintenance needs before they occur, thereby preventing costly downtime and extending the lifespan of the infrastructure. In a research by Soothar et al. [68], ML & DL techniques were used to classify six types of optical network faults and evaluate the accuracy and delay of various models. The CNN-LSTM hybrid model achieved the highest accuracy of 99%, and the proposed fault prevention algorithm enhances network reliability.

A study by Liu et al. [69] implemented an ML model for real-time analysis of DFOS data to monitor both uniform and non-uniform pipeline corrosion. The model automatically detects corrosion and quantifies it based on DFOS data. Evaluated through laboratory experiments, the approach demonstrates high accuracy, with an F1 score of 0.986 for corrosion detection and an R^2 value of 0.953 for corrosion quantification. This method improves pipeline corrosion monitoring by automating the interpretation of DFOS data, making the process more efficient and robust.

AI is instrumental in monitoring the security and intrusion detection functions of OFSNs. With their high sensitivity and extensive coverage, fiber sensors are commonly employed for perimeter security and intrusion detection, making them highly effective in safeguarding large areas. A study by Yang et al. [70] developed a distributed fiber intrusion identification algorithm using DL to enhance detection accuracy. The algorithm integrates a CNN to extract spatial features and a BP neural network to capture temporal features from traffic data. By stacking CNN+BPNN modules, both spatial and temporal characteristics are effectively extracted. The results show a recognition rate of 93% for detecting common intrusion events.

All the above studies indicate that the integration of AI with OFSNs represents a paradigm shift in sensor technology, enabling more accurate, reliable, and cost-effective solutions for a wide range of applications. From enhancing data processing and signal analysis to optimizing sensor design and improving security measures, AI-driven innovations are unlocking new possibilities for the advancement of OFSNs.

5 Applications of AI Integrated OFSN and Future Scope of the Work

AI-driven optimization has significantly enhanced real-world applications of OFSNs across various industries (Fig. 4). For instance, in the oil and gas industry, DOFS are installed for pipeline monitoring [71]. AI algorithms analyze this data to predict potential leaks or structural issues, allowing for predictive maintenance, which reduces downtime and mitigates environmental risks. Similarly, in structural health monitoring [72], optical fiber sensors embedded in bridges and buildings provide real-time data on strain, temperature, and vibration. AI-powered systems analyze these signals to assess structural integrity, enabling timely interventions that extend the life of critical infrastructure and ensure safety.

In the aerospace sector [73], optical fiber sensors have been integrated into aircraft components to continuously monitor vital parameters. AI-driven anomaly detection algorithms process this data to identify mechanical issues or fatigue before they become critical. This improves flight safety through preemptive maintenance. OFSNs are also pivotal in environmental monitoring [74], where they detect pollutants in air, water, and soil. AI systems analyze the data from these sensors to identify abnormal pollutant levels or environmental changes, supporting pollution control and sustainable resource management.

In industry, AI-driven optimization enhances the performance of OFSNs in manufacturing machine diagnostics [75]. These sensors track operational parameters

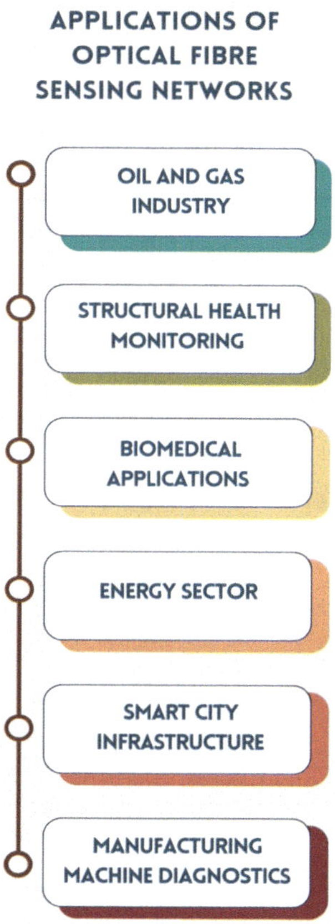

Fig. 4 Applications of optical fiber sensors in various domains

throughout production. AI algorithms analyze this data to enhance production efficiency, reduce defects, and ensure consistent product quality. AI has also transformed optical wireless communications [76], where optical fiber sensor networks along with DL algorithms can be applied for optical wireless transceiver design.

AI-driven optical fibers also find many useful biomedical applications, such as monitoring physiological parameters such as heart rate, blood glucose levels, and oxygen saturation [77]. AI processes this sensor data to provide real-time health insights, facilitating remote patient monitoring and enabling early detection of medical complications, which supports personalized healthcare delivery and improves patient outcomes. In environmental sensing networks, AI-driven optimization plays a key role in natural disaster early warning systems [78]. When these networks are deployed in disaster-prone areas, they detect seismic vibrations and ground movements. This allows AI methods to provide early warnings for earthquakes or landslides, thus contributing to public safety and disaster preparedness.

In smart cities [79], OFSNs are integrated into urban infrastructure, where AI analyzes this data to assess structural integrity, detect leaks, and optimize resource use, making urban infrastructure more efficient and reducing maintenance costs while supporting sustainable urban development. In the energy sector [80], OFSNs monitor power grids, tracking parameters like temperature, strain, and electrical current. AI algorithms can also predict equipment failures and optimize energy distribution.

Thus, from the above study, it is evident that across a variety of diverse sectors, AI-driven OFSNs have delivered improvements in operational efficiency, safety, and sustainability. By leveraging AI for predictive analytics, anomaly detection, and real-time optimization, industries can address complex challenges and drive innovation. The ongoing advancement of AI-driven OFSNs continues to unlock new potential and fosters smarter, more resilient solutions in a variety of domains.

References

1. C. Pendão, I. Silva, Optical fiber sensors and sensing networks: overview of the main principles and applications. Sens. **22**(19), 7554 (2022)
2. I. Ashry, Y. Mao, B. Wang, F. Hveding, A.Y. Bukhamsin, T.K. Ng, B.S. Ooi, A review of distributed fiber–optic sensing in the oil and gas industry. J. Light. Technol. **40**(5), 1407–1431 (2022)
3. E.J. Friebele, C.G. Askins, A.B. Bosse, A.D. Kersey, H.J. Patrick, W.R. Pogue, S.T. Vohra et al., Optical fiber sensors for spacecraft applications. Smart Mater. Struct. **8**(6), 813 (1999)
4. E. Ip, F. Ravet, H. Martins, M.F. Huang, T. Okamoto, S. Han, C. Narisetty, J. Fang, Y.K. Huang, M. Salemi, E. Rochat, F. Briffod, A. Goy, M. del Rosario Fernández-Ruiz, M.G. Herráez, Using global existing fiber networks for environmental sensing. Proc. IEEE **110**(11), 1853–1888 (2022)
5. R. Correia, S. James, S.W. Lee, S.P. Morgan, S. Korposh, Biomedical application of optical fibre sensors. J. Opt. **20**(7), 073003 (2018)
6. A. Loayssa, Optical fiber sensors for structural health monitoring, in *New Developments in Sensing Technology for Structural Health Monitoring*. (Springer Berlin Heidelberg, Berlin, Heidelberg, 2011). pp. 335–358

7. E. Ip, J. Fang, Y. Li, Q. Wang, M.F. Huang, M. Salemi, Y.K. Huang, Distributed fiber sensor network using telecom cables as sensing media: technology advancements and applications. J. Opt. Commun. Netw. **14**(1), A61–A68 (2022)
8. S. Ahmed Hazim, A. Faleh Mahmood, Developments in optical fiber network fault detection methods: an extensive analysis. Int. J. Comput. Digit. Syst. **16**(1), 1–12 (2024)
9. K. Abdelli, J.Y. Cho, F. Azendorf, H. Griesser, C. Tropschug, S. Pachnicke, Machine-learning-based anomaly detection in optical fiber monitoring. J. Opt. Commun. Netw. **14**(5), 365–375 (2022)
10. Q. Wang, G. Farrell, W. Yan, Investigation on single-mode–multimode–single-mode fiber structure. J. Light. Technol. **26**(5), 512–519 (2008)
11. L. Shen, C. Teng, Z. Wang, H. Bai, S. Kumar, R. Min, Semiconductor multimaterial optical fibers for biomedical applications. Biosens. **12**(10), 882 (2022)
12. R. Min, Z. Liu, L. Pereira, C. Yang, Q. Sui, C. Marques, Optical fiber sensing for marine environment and marine structural health monitoring: a review. Opt. & Laser Technol. **140**, 107082 (2021)
13. H. Zhang, Y. Gong, D. Jia, T. Liu, Y. Zhang, Robustness analysis based on optical fiber sensor networks topology. IEEE Sens. J. **15**(3), 1388–1394 (2014)
14. A. Banerjee, Y. Park, F. Clarke, H. Song, S. Yang, G. Kramer, K. Kim, B. Mukherjee, Wavelength-division-multiplexed passive optical network (WDM-PON) technologies for broadband access: a review. J. Opt. Netw. **4**(11), 737–758 (2005)
15. R.S. Tessinari, M.H.M. Paiva, M.E. Monteiro, M.E.V. Segatto, A.S. Garcia, G.T. Kanellos, R. Nejabati, D. Simeonidou, On the impact of the physical topology on the optical network performance, in *2018 IEEE British and Irish Conference on Optics and Photonics (BICOP)* (IEEE, 2018). pp. 1–4
16. D.M. Spirit, A.D. Ellis, P.E. Barnsley, Optical time division multiplexing: systems and networks. IEEE Commun. Mag. **32**(12), 56–62 (1994)
17. J. Yu, N. Chi, *Digital Signal Processing in High-Speed Optical Fiber Communication Principle and Application.* (Springer Nature, 2020)
18. S. Deepa, B. Das, Interrogation techniques for π-phase-shifted fiber Bragg grating sensor: a review. Sens. Actuators A: Phys. **315**, 112215 (2020)
19. M. Elsherif, A.E. Salih, M.G. Muñoz, F. Alam, B. AlQattan, D.S. Antonysamy, M.F. Zaki, A.K. Yetisen, S. Park, T.D. Wilkinson, H. Butt, Optical fiber sensors: working principle, applications, and limitations. Adv. Photonics Res. **3**(11), 2100371 (2022)
20. M.M. Eid, Optical fiber sensors: review of technology and applications. Indones. J. Electr. Eng. Comput. Sci. **25**(2), 1038–1046 (2022)
21. R. Kashyap, *Fiber Bragg Gratings.* (Academic press, 2009)
22. S.K. Ibrahim, J.A. O'Dowd, V. Bessler, D.M. Karabacak, J.M. Singer, Optimization of fiber Bragg grating parameters for sensing applications, in *Fiber Optic Sensors and Applications XIV*, vol. 10208 (SPIE, 2017). pp. 163–173
23. A. Martinez-Rios, D. Monzon-Hernandez, G. Salceda-Delgado, I. Torres-Gomez, *Long Period Fibre Gratings.* (INTECH Open Access Publisher, 2012)
24. X.W. Zhao, Q. Wang, Mini review: recent advances in long period fiber grating biological and chemical sensors. Instrum. Sci. & Technol. **47**(2), 140–169 (2019)
25. T.K. Yadav, R. Narayanaswamy, M.H. Abu Bakar, Y.M. Kamil, M.A. Mahdi, Single mode tapered fiber-optic interferometer based refractive index sensor and its application to protein sensing. Opt. Express **22**(19), 22802–22807 (2014)
26. T. Yoshino, K. Kurosawa, K. Itoh, T. Ose, Fiber-optic fabry-perot interferometer and its sensor applications. IEEE Trans. Microw. Theory Tech. **30**(10), 1612–1621 (1982)
27. Q. Yu, X. Zhou, Pressure sensor based on the fiber-optic extrinsic fabry-perot interferometer. Photonic Sens. **1**, 72–83 (2011)
28. K. Thurner, P.F. Braun, K. Karrai, Fabry-Pérot interferometry for long range displacement sensing. Rev. Sci. Instrum. **84**(9) (2013)
29. L. Li, L. Xia, Z. Xie, D. Liu, All-fiber mach-zehnder interferometers for sensing applications. Opt. Express **20**(10), 11109–11120 (2012)

30. U. Sharma, X. Wei, Fiber optic interferometric devices, in *Fiber Optic Sensing and Imaging* (Springer New York, New York, NY, 2013). pp. 29–53
31. F. Xie, J. Ren, Z. Chen, Q. Feng, Vibration-displacement measurements with a highly stabilised optical fiber Michelson interferometer system. Opt. & Laser Technol. **42**(1), 208–213 (2010)
32. L.R. Jaroszewicz, Angular velocity measurements in sagnac interferometer, in *Optical Velocimetry*, vol. 2729 (SPIE, 1996). pp. 191–196
33. N.A. Aljbar, B. R. Mahdi, A.H. Khalid, A.H. Attallah, F.S. Abdulwahid, A.J. Haider, Enhanced surface plasmon resonance (SPR) fiber optic sensor for environmental monitoring: a coreless fiber–based design. Plasmonics. 1–10 (2024)
34. E. Benito-Peña, M.G. Valdés, B. Glahn-Martínez, M.C. Moreno-Bondi, Fluorescence based fiber optic and planar waveguide biosensors. A review. Anal. Chim. Acta **943**, 17–40 (2016)
35. W.S. Klubben, S.L. Logunov, E.J. Fewkes, J. Mooney, P.M. Then, P.G. Wigley, H. Schreiber, K. Matias, C.J. Wilson, M. Ocampo, Novel light diffusing fiber for use in medical applications. in *Optical Fibers and Sensors for Medical Diagnostics and Treatment Applications XVI*, vol. 9702 (SPIE, 2016), pp. 234–239
36. R.A. Perez-Herrera, M. Lopez-Amo, Fiber optic sensor networks. Opt. Fiber Technol. **19**(6), 689–699 (2013)
37. A. Venketeswaran, N. Lalam, J. Wuenschell, P.R. Ohodnicki Jr., M. Badar, K.P. Chen, P. Lu, Y. Duan, B. Chorpening, M. Buric, Recent advances in machine learning for fiber optic sensor applications. Adv. Intell. Syst. **4**(1), 2100067 (2022)
38. K. Dey, N. Vangety, S. Roy, Machine learning approach for simultaneous measurement of strain and temperature using FBG sensor. Sens. Actuators A: Phys. **333**, 113254 (2022)
39. Y. Han, S. Yu, M. Li, J. Yang, W. Gu, An SVM-based detection for coherent optical APSK systems with nonlinear phase noise. IEEE Photonics J. **6**(5), 1–10 (2014)
40. E. Yigit, E. Hayber, U. Aydemir, U. Ozkaya, Machine learning-based estimation of static and dynamic pressure responses of specially designed diaphragm based fabry–perot interferometer, in *Proceedings 2nd International Conference on Engineering and Applied Natural Sciences* (2022), pp. 1–4
41. A. Usman, N. Zulkifli, M.R. Salim, K. Khairi, Fault monitoring in passive optical network through the integration of machine learning and fiber sensors. Int. J. Commun. Syst. **35**(9), e5134 (2022)
42. S.P. Abdula, M.J. Llagas, A.M. Fernandez, E. Arboleda, *Machine Learning Applications for Fault Tracing and Localization in Optical Fiber Communication Networks: A Review* (2024)
43. M.R. Mahani, Y. Rahimof, S. Wenzel, I. Nechepurenko, A. Wicht, Data-efficient machine learning algorithms for the design of surface Bragg gratings. ACS Appl. Opt. Mater. **1**(8), 1474–1484 (2023)
44. D. Wang, M. Zhang, Z. Cai, Y. Cui, Z. Li, H. Han, M. Fu, B. Luo, Combatting nonlinear phase noise in coherent optical systems with an optimized decision processor based on machine learning. Opt. Commun. **369**, 199–208 (2016)
45. N.H. Al-Ashwal, K.A. Al Soufy, M.E. Hamza, M.A. Swillam, Deep learning for optical sensor applications: a review. Sens. **23**(14), 6486 (2023)
46. Z. Tan, S. Yu, W. Yang, Y. Ren, R. Pan, X. Yu, A mach-zehnder-based optical fiber sensor enables multi-directional and wide-range curvature sensing with the assistance of CNN. Opt. Fiber Technol. **84**, 103766 (2024)
47. C. Wei, Q. Liu, Y. Wang, D. Zhu, J. Shi, D. Lin, Accurate strain sensing with fiber-optic fabry-perot sensors based on CNN-LSTM Model. IEEE Sens. J. (2024)
48. Y. Mei, S. Zhang, Z. Cao, T. Xia, X. Yi, Z. Liu, High accuracy pressure sensing with Sagnac interferometry based on deep learning approach, in *2022 IEEE 24th International Workshop on Multimedia Signal Processing (MMSP)* (IEEE, 2022), pp. 1–5
49. T. Yamaguchi, H. Kawashima, H. Matsuda, Y. Shinoda, Experimental demonstration of peak wavelength measurement of multiplexing fiber bragg gratings using convolutional neural network. IEEE Sens. J. **23**(9), 9343–9352 (2023)
50. H. Liu, J. Xia, J. Xin, H. Zhao, Y. Ni, Wavelength detection technique of overlapping spectra in the serial WDM FBGs by convolutional neural network. Opt. Fiber Technol. **75**, 103206 (2023)

51. L.V. Nguyen, C.C. Nguyen, G. Carneiro, H. Ebendorff-Heidepriem, S.C. Warren-Smith, Sensing in the presence of strong noise by deep learning of dynamic multimode fiber interference. Photonics Res. **9**(4), B109–B118 (2021)
52. J. Ali, A. Almaiman, M.A. Esmail, A.M. Ragheb, E.M. Almohimmah, H. Bryhni, S.A. Alshebeili, Dual-stage deep learning for sangac optical fiber sensing multi-event detection and localization. Opt. & Laser Technol. **179**, 111295 (2024)
53. S. Li, S. Ren, S. Chen, B. Yu, Improvement of fiber Bragg grating wavelength demodulation system by cascading generative adversarial network and dense neural network. Appl. Sci. **12**(18), 9031 (2022)
54. L. Wang, H. Liu, Z. Pan, Y. Xu, D. Fan, C. Zhou, Y. Li, Temperature demodulation for optical fiber FP sensor based on DBNs with ensemble learning. Opt. & Laser Technol. **162**, 109275 (2023)
55. B. Karanov, M. Chagnon, F. Thouin, T.A. Eriksson, H. Bülow, D. Lavery, P. Bayvel, L. Schmalen, End-to-end deep learning of optical fiber communications. J. Light. Technol. **36**(20), 4843–4855 (2018)
56. E. Reyes-Vera, A. Valencia-Arias, V. García-Pineda, E.F. Aurora-Vigo, H. Alvarez Vásquez, G. Sánchez, Machine learning applications in optical fiber sensing: a research agenda. Sens. **24**(7), 2200 (2024)
57. Y.C. Manie, J.W. Li, P.C. Peng, R.K. Shiu, Y.Y. Chen, Y.T. Hsu, Using a machine learning algorithm integrated with data de-noising techniques to optimize the multipoint sensor network. Sens. **20**(4), 1070 (2020)
58. D. Wang, C. Zhang, W. Chen, H. Yang, M. Zhang, A.P.T. Lau, A review of machine learning-based failure management in optical networks. Sci. China Inf. Sci. **65**(11), 211302 (2022)
59. C. Quach, S. Vazquez, A. Tessler, J. Moore, E. Cooper, J. Spangler, Structural anomaly detection using fiber optic sensors and inverse finite element method, in *AIAA Guidance, Navigation, and Control Conference and Exhibit* (2005), p. 6357
60. H. Li, K. Li, H. Li, F. Meng, X. Lou, L. Zhu, Recognition and classification of FBG reflection spectrum under non-uniform field based on support vector machine. Opt. Fiber Technol. **60**, 102371 (2020)
61. B. Raju, R. Kumar, S. Dhanalakshmi, Design and implementation of tilted FBG for concurrent temperature and humidity measurement using machine learning. Opt. Fiber Technol. **82**, 103630 (2024)
62. R. Martinek, R. Kahankova, M. Fajkus, J. Nedoma, M. Novak, J. Jargus, Advanced methods for fiber-optic sensor signal processing, in *Fiber Optic Sensors and Applications XV*, vol. 10654 (SPIE, 2018), pp. 334–340
63. Z. Wang, A. Yang, P. Guo, P. He, OSNR and nonlinear noise power estimation for optical fiber communication systems using LSTM based deep learning technique. Opt. Express **26**(16), 21346–21357 (2018)
64. Hou, L, T. Jiang, T. Yu, C. Cao, X. Tu, J. Zhang, J. Pan, S. Wang, N. Zhou, N. Yao, L. Zhang, Tapered optical fiber enabled distributed sensors with high spatial resolution by deep learning. Adv. Opt. Mater. 2303118 (2024)
65. S. Hassani, U. Dackermann, A systematic review of optimization algorithms for structural health monitoring and optimal sensor placement. Sens. **23**(6), 3293 (2023)
66. Y. Dogan, R. Katirci, İ Erdogan, E. Yartasi, Artificial neural network based optimization for Ag grated D-shaped optical fiber surface plasmon resonance refractive index sensor. Opt. Commun. **534**, 129332 (2023)
67. G. Kaur, R.S. Kaler, N. Kwatra, On the optimization of fiber bragg grating optical sensor using genetic algorithm to monitor the strain of civil structure with high sensitivity. Opt. Eng. **55**(8), 087103–087103 (2016)
68. K.K. Soothar, Y. Chen, A.H. Magsi, C. Hu, H. Shah, Optimizing optical fiber faults detection: a comparative analysis of advanced machine learning approaches. Comput., Mater. & Contin. **79**(2) (2024)
69. Y. Liu, X. Tan, Y. Bao, Machine learning-assisted intelligent interpretation of distributed fiber optic sensor data for automated monitoring of pipeline corrosion. Meas. **226**, 114190 (2024)

70. N. Yang, Y. Zhao, Distributed optical fiber intrusion identification algorithm based on deep learning, in *2023 2nd International Conference on Artificial Intelligence and Autonomous Robot Systems (AIARS)* (IEEE, 2023), pp. 458−462
71. W.A. Altabey, Z. Wu, M. Noori, H. Fathnejat, Structural health monitoring of composite pipelines utilizing fiber optic sensors and an AI-based algorithm—a comprehensive numerical study. Sens. **23**(8), 3887 (2023)
72. U.M.N. Jayawickrema, H.M.C.M. Herath, N.K. Hettiarachchi, H.P. Sooriyaarachchi, J.A. Epaarachchi, Fibre-optic sensor and deep learning-based structural health monitoring systems for civil structures: a review. Meas. **199**, 111543 (2022)
73. F. Kosova, Ö. Altay, H.Ö. Ünver, Structural health monitoring in aviation: a comprehensive review and future directions for machine learning. Nondestruct. Test. Eval. 1−60 (2024)
74. L.S. Goh, K. Onodera, M. Kanetsuna, K. Watanabe, N. Shinomiya, Constructing an optical fiber sensor network for natural environment remote monitoring, in *The 17th Asia Pacific Conference on Communications* (IEEE, 2011), pp. 208−212
75. M. Ferrario, M. Mattarei, P. Boffi, M. Martinelli, A software-defined coherent fiber optic sensor for manufacturing machine diagnostic, in *2016 IEEE Sensors Applications Symposium (SAS)* (IEEE, 2016), pp. 1−5
76. H. Lee, S.H. Lee, T.Q. Quek, I. Lee, Deep learning framework for wireless systems: applications to optical wireless communications. IEEE Commun. Mag. **57**(3), 35–41 (2019)
77. C. Perezcampos Mayoral, J. Gutiérrez Gutiérrez, J.L. Cano Pérez, M. Vargas Treviño, I.B. Gallegos Velasco, P.A. Hernández Cruz, R.T. Rosas, L.T. Carrillo, J.A. Ríos, E.L. Apreza, R. Rojas Laguna, Fiber optic sensors for vital signs monitoring. A review of its practicality in the health field. Biosens. **11**(2), 58 (2021)
78. P.D. Hernández, J.A. Ramírez, M.A. Soto, Deep-learning-based earthquake detection for fiber-optic distributed acoustic sensing. J. Light. Technol. **40**(8), 2639–2650 (2021)
79. N.R. Kapoor, A. Kumar, H.C. Arora, A. Kumar, Structural health monitoring of existing building structures for creating green smart cities using deep learning, in *Recurrent Neural Networks* (CRC Press, 2022). pp. 203−232
80. Q. Chai, Y. Luo, J. Ren, J. Zhang, J. Yang, L. Yuan, G.D. Peng, Review on fiber-optic sensing in health monitoring of power grids. Opt. Eng. **58**(7), 072007–072007 (2019)

Smart Infrastructure: Integrating IOT, AI, and Optical Fiber Sensor

Arvind Sharma, Kavita Rani Segwal, Umang Sharma, Sugandha Gupta, and Deepika Rathore

Abstract The integration of crucial devices such as artificial intelligence (AI), Internet of Things (IoT), and optical fiber sensors (OFS), considering the urbanization problem, has enhanced the quality of life, making it more sustainable and ecologically friendly. Rapidly changing technologies set new benchmarks for urban development and make cities more efficient and resilient. IoT devices connect the physical devices to enable live monitoring in case of energy management, traffic control, network communication, and structural health. Vast data has been generated and analyzed using AI, improving operations and providing predictive information. Physical parameters monitoring uses OFS that are sensitive and unaffected by ecological disturbances. OFS integration with an IoT network offers an AI system with accurate data, ensuring live tracking. This integration minimizes environmental impact, reduces downtime, increases life-useful benefits, and improves resource management.

Keywords Optical fibre sensors (OFS) · Internet of things (IoT) · Artificial intelligence (AI)

1 Introduction

Smart sensors and IoT have been transformed entirely by incorporating the best technologies into sensor networks and dialogue systems. Optical technologies are a critical component in advancing sensor communication networks' performance and reliability, and they explore the role of determining the outcomes of complex systems and related applications [1]. Hence, integrating IoT into optical sensors,

A. Sharma (✉) · K. R. Segwal · U. Sharma · S. Gupta
Department of Applied Science (Physics), Bhagwan Parshuram Institute of Technology, Delhi, India
e-mail: arvindsharma1900@gmail.com

D. Rathore
Department of Applied Science (Mechanical), Bhagwan Parshuram Institute of Technology, Delhi, India

V. Bhardwaj et al. (eds.), *Optical Fiber Sensors and AI*, Progress in Optical Science and Photonics 34, https://doi.org/10.1007/978-981-96-6158-9_8

optical communication systems, and optical networks provides seamless communication, unique record acquisition, and green processing [2]. Intelligent transportation could improve traffic efficiency and safety if the road monitoring system were reliable. Based on its three drivers, decarbonization, digitalization, and decentralization, a smart city's electrical distribution system, or smart grid, receives the critical service made possible by technology for its energy resources [3–5]. Smart cities and infrastructure that led to a reduction in carbon footprint only filled the growing demand for sustainable solutions to electricity-related issues, such as effective monitoring and management of electricity throughout its entire process, including generation, transmission, distribution, and storage. Large-scale combinations of networks with interdependent interactions are also considered parts of smart cities [5]. Integrating IoT and OFS has many benefits, such as reducing several sensors, which improves precise measurement and increases reliability. Data collection, transmission, and processing are transformed by optical wireless sensor networks, distributed optical sensing, and passive optical sensors. Incorporating visible light communication and energy harvesting capability enhanced the functionality and efficiency of IoT devices [6–8].

1.1 Importance of IoT Integration

For a sustainable society to build a smart infrastructure, the fifth-generation wireless technology for 6G uses IoT devices that allow real-time data control and communication, enabling harvesting and analysis. This integration is advantageous because of:

i. **Enhanced Connectivity**: IoT achieves seamless communication between interface elements, sensors, and systems through data exchange and communication [9].
ii. **Data Collection and Analysis**: IoT devices collect essential data from multiple sources to get decision-making information and integrate it to enhance resource utilization, reduce costs, and achieve better performance [9].
iii. **Automation and Efficiency**: Integrating IoT can automate processes, increasing efficiency and productivity by reducing human intervention and potential errors [9].
iv. **Smart Logistics**: By optimizing communication between IoT devices, logistics architecture for businesses can be improved with complete automation and seamless warehouse communication [9].
v. **User-Centric Services**: By developing user-centric wireless services, which are crucial for smart cities, smart logistics, and precision agriculture, IoT integration improves the user experience [10].
vi. **Security and Monitoring**: IoT devices may be used for security, monitoring different parts of the infrastructure and promptly informing stakeholders to improve the safety and security of the smart infrastructure [10].

1.2 Importance of OFS Integration

Monitoring the condition of concrete structures, specially reinforced with textile-reinforced concrete OFS integration, becomes essential and with benefits, such as resistance to high temperatures, multiplexing capability, and immunity to EM interference. Hence, it is ideal for detecting issues of concrete buildings. By integrating OFS, we may establish a network of sensors along a single fiber, conduct routine inspections, and identify structural deterioration early [11]. The benefits of OFS are as follows:

i. **High Sensitivity and Accuracy**: It precisely measures critical infrastructure.
ii. **Data Transmission**: Data is transmitted for distances without major loss.
iii. **Durability and Reliability**: It can resist harsh environmental conditions, confirming long-term reliability.
iv. **Live monitoring**: It allows for informed decisions.

1.3 Importance of AI Integration

i. **Enhanced performance**: Integrating AI in smart grids significantly enhances grid reliability, load prediction accuracy, fault detection capabilities, and cybersecurity against potential cyber-attacks.
ii. **Efficient Data Handling**: AI technologies facilitate intelligent grid operations and well-informed decision-making by effectively handling and evaluating large-scale data.
iii. **Future-Proofing Strategies**: Integrating artificial intelligence (AI) solutions into an already-existing smart grid improves its capacity to foresee new issues and adjust to emerging technologies, guaranteeing sustainability and resilience in the evolving energy landscape.
iv. **Enhanced Decision-Making**: Optical sensors enhanced by AI provide better decision-making abilities that can improve efficiency and precision when analyzing data.
v. **Local Intelligence Development**: This development is facilitated by integrating artificial intelligence into optical sensors, which enable real-time data acquisition and processing within a sensor network.
vi. **Global Intelligence Integration**: Global intelligence integration is accomplished by linking AI-driven sensors to a central network management center. This advances grid monitoring and control capabilities, elevating them to a higher level [12].

2 Role of OFS Integration with AI and IoT in Smart Infrastructure

The advancement of technologies in IoT, AI, and OFS has transformed cities and set new benchmarks for urban development. It shows how this technology integration creates sustainable urban systems with examples. The findings revealed that the grown technologies promote productivity and improve quality of life. At the same time, integration revolutionizes quality and sets new benchmarks in smart infrastructure. The emergence of smart infrastructure is linked with eco-friendly development and optimized resource allocation. Therefore, emphasizing technology integration is necessary to enhance the quality of life for city dwellers and ensure proper city functionality [13, 14]. Much research on smart infrastructure with time integrates modern technologies into traditional systems [15], like transportation, energy, and water supply [16]. This integration allows live monitoring of smart infrastructure and better management. The key driving parameters for the evolution of smart infrastructure are as follows: IoT connects physical objects via electronics, software, and sensors, allowing live monitoring between communication networks and making the urban system more efficient and responsive [17]. Cloud computing is required to process and store such massively generated data. This will help the cities to make informed decisions and optimize operations [13, 14, 18]. Concurrently, artificial intelligence is used to analyze the data. It provides a predictive and automated process and helps with proper energy utilization and traffic flow, improving efficiency and sustainability [19]. Smart infrastructure capabilities are improved using big data analytics through large datasets, helping cities with trends, patterns, and opportunities [17, 20].

Meanwhile, OFS becomes an essential parameter in smart infrastructure development, providing live monitoring for health infrastructure. Their integration with other smart systems helps us monitor structural health, detect anomalies, and safeguard infrastructure integrity. Optical fibre sensors actively help by providing on-time information about physical parameters, such as strain, temperature, and vibrations, enhancing the smart infrastructure's ability to prevent failure. Hence, its accurate and reliable measurement makes the smart infrastructure indispensable for confirming longevity and resilience and can withstand harsh environmental conditions. The continuous evolution of optical fiber sensor technology shows potential for monitoring and managing smart infrastructures [21, 22], see Fig. 1. Constant research efforts and growing technologies have shaped the smart infrastructure from zero to the pro level. Therefore, integrated technologies create feasible and livable urban environments for their residents.

The image illustrates a conceptual diagram of a 'Smart City' ecosystem, showcasing various interconnected systems that utilize IoT (Internet of Things) technology to create a more efficient, sustainable, and liveable urban environment. Each component can be explained as follows:

- Smart Retail: Retail systems that use IoT to enhance customer experience, optimize inventory, and streamline operations like automated checkout systems,

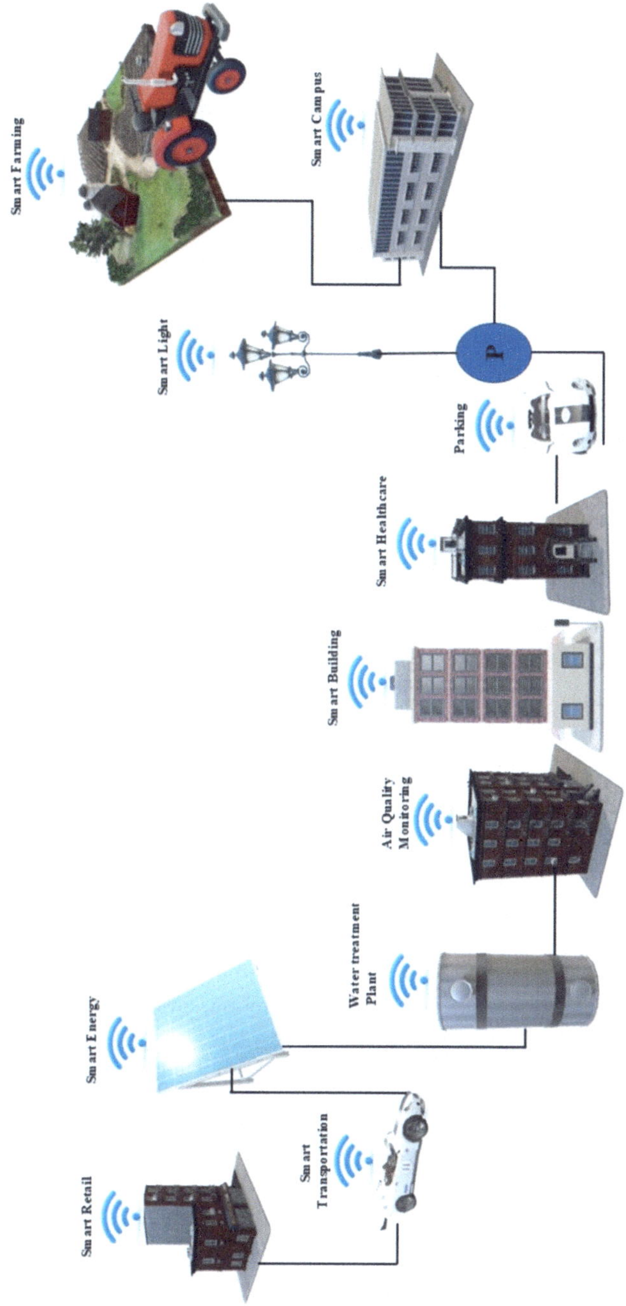

Fig. 1 Pictorial representation of smart infrastructure integration

personalized shopping experiences based on data analytics, and real-time stock management.

- Smart Transportation: It integrates IoT with transportation systems to improve traffic flow, reduce emissions, and enhance safety, such as EV charging stations, traffic management systems, and autonomous vehicles.
- Smart Energy: It uses IoT to manage energy production, distribution, and consumption more efficiently, such as smart grids, renewable energy sources like wind and solar power, and energy-saving home appliances.
- Water Treatment and Management: It utilizes IoT for efficient water distribution, quality monitoring, and wastewater treatment, like automated irrigation systems, real-time water quality sensors, and leak detection systems.
- Air Quality Monitoring: Systems that monitor pollution levels and air quality in real-time, helping to create healthier environments like sensors placed throughout the city that track pollutants and provide data for public health interventions.
- Smart Building: Buildings equipped with IoT technology to optimize energy usage, enhance security, and improve comfort, like automated climate control, smart lighting, and security systems.
- Smart Healthcare: Health is linked to telemedicine, smart inhalers, wearable health devices such as smartwatches (Fitbit, Apple watch), remote patient monitoring, and data-driven patient care, all part of the smart healthcare ecosystem. The potential for IoT in healthcare is vast, from personalized treatment plans to predictive analytics that anticipate health issues before they escalate [23].
- Parking System: Smart parking uses IoT-enabled sensors to monitor real-time parking occupancy. Its benefits include convenience for drivers, reduced fuel consumption, and efficient parking management like San Francisco's SF park program, Singapore's smart parking guidance, and Barcelona's multifunctional parking app [24].
- Smart Street Light: IoT-enabled streetlights adjust brightness based on real-time conditions. The advantages are energy efficiency, cost savings, and improved safety, such as Amsterdam's dimming lights, Los Angeles' adaptive lighting, and Singapore's air quality monitoring streetlights [25]. These lights adapt to ambient conditions, brighten when needed, and conserve energy when streets are empty. The streetlights are equipped with sensors, communication, and management modules to detect changes in ambient light, motion, and other environmental factors.
- Smart Farming: Smart farming leverages IoT to optimize agricultural processes. Sensors collect data on soil moisture, crop health, and more. It aids efficiency, productivity, and sustainability like precision planting, connected cows, and autonomous tractors [26].

2.1 Execution of Smart Integration

Figure 2 illustrates the steps involved in the implementation of Smart Integration; details for each step are outlined as follows:

Step 1: Install smart infrastructure devices like network, AI, and IoT. **Step 2**: For Physical parameter monitoring, install an OFS sensor. **Step 3**: Collect real-time data from deployed sensors. **Step 4**: Use IoT to send sensor data to the cloud server. **Step 5**: The processor processes the received data. **Step 6**: Filter the raw data by removing noise for further analysis. **Step 7**: Preprocessed data analyzed using AI or Machine learning. **Step 8**: Model identifies the data analytic that causes the potential inadequacy. **Step 9**: Store the process and analysis data for future study in the database. **Step 10**: Generate live information, altering for a major occurrence. **Step 11**: Shows information and warnings on a user-friendly dashboard. **Step 12**: The system makes computerized adjustments or notifies the operator. **Step 13**: Upgrading the model with continuous feedback to enhance the precision.

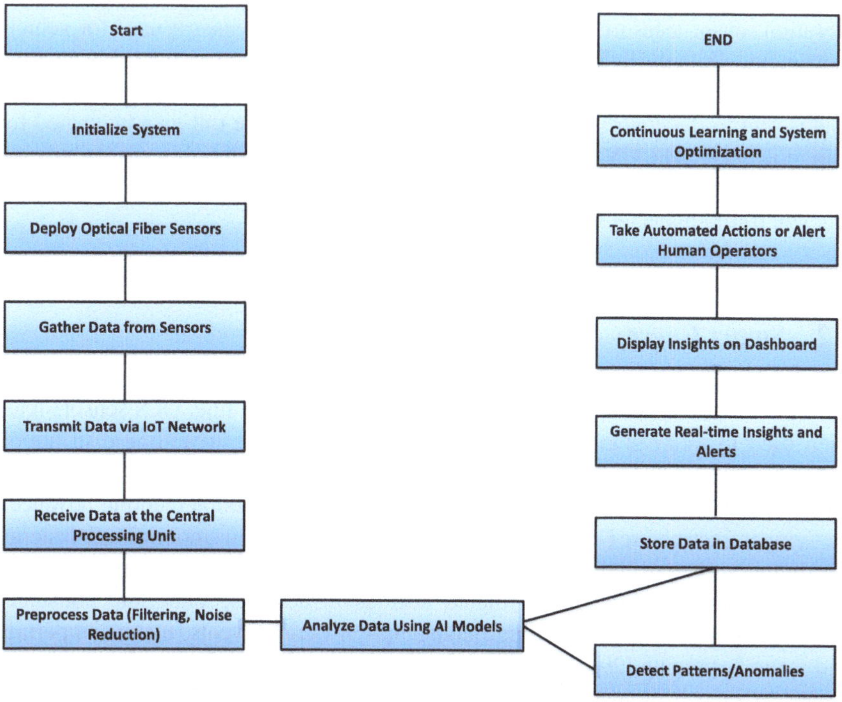

Fig. 2 Flow chart for smart integration

3 Applications, Challenges, and Solutions

Integrating IoT, AI, and optical fiber sensors creates a powerful collaboration, enabling the development of ingenious infrastructure. IoT devices collect huge amounts of data from various sensors, and AI analyses these data to generate practical findings. Optical fiber sensors provide precise, real-time measurements.

3.1 *Smart Bridge*

Innovative bridges with IoT devices and optical fiber sensors continuously monitor structural integrity, detecting stress, strain, and potential faults. AI algorithms analyze the data to predict the maintenance required for the safety and longevity of the bridge [2]. Darshan et al. reported [27] that the moveable bridge enhances traffic flow by preserving the canal access. The system has many components, such as flood monitoring, a water indicator for measuring bending, and a flex sensor. The vibration sensor measures the vibration of a bridge; an IR sensor is used for vehicle movement, while the light sensor detects whether it is daytime or night. Meanwhile, the ultrasonic sensor is used to track the ship. After tracking till 6 km from the bridge, data is sent to the control room to be analyzed, and then the toll gate is opened for ship movement. The UV sensor repeats the same process as the ship passes but now closes the toll gate. Hence, the primary objective of tracking bridge conditions and efficient traffic management is fulfilled.

3.2 *Smart Energy Grids*

For key devices such as detecting faults in the transmission line, OFS sensors have been used to monitor energy consumption IoT and AI to streamline distribution and avoid failure. The smart grid development in India emphasizes the integration of AI and IoT to increase reliability and productivity. Implementing advanced metering infrastructure (AMI) systems is feasible, and real-time monitoring of energy consumption patterns, alarms, and various communication technologies is possible [28]. The AMI system has facilitated utilities in detecting abnormal consumer behavior in real-time, such as high consumption during events, meter tampering, and meter bypassing [29]. It has helped to recognize unbalanced power supply and variations in voltage at the consumer end, leading to improved system efficiency [29]. The Smart Grid Control Centre established in Puducherry is the first of its kind in India, enabling real-time transformer-wise distribution energy audits and outage management through technologies such as distribution transformer monitoring units (DTMU) and fault passage indicators (FPI) [28]. AI and IoT technologies are essential for the energy sector modernization, meeting consumers' growing demands, and securing power supply at minimal cost. By harnessing these technologies, the power engineering community has been better equipped to face challenges, enhance operational efficiency, and enable a sustainable and digitally enabled power ecosystem [28–30].

3.3 Smart Transportation Systems

Strain and vibration on roads and railways can be detected using OFS sensors. IoT devices use proper traffic management to reduce traffic jams with the help of AI algorithms, enabling better road safety. Concurrently, a recent study on the "Intelligent Solutions for Road Safety through Technology and Engineering" project was conducted jointly by CSIR Central Road Research Institute and Ashok Leyland Institute for Driving Training & Research [31]. The main principle is road safety through innovative solutions. The Project integrates technology and engineering solutions, including artificial intelligence for driver assistance and collision avoidance systems. With consortium partners and oversight from the Nagpur Municipal Corporation, project iRASTE aims to pioneer road safety interventions based on leading indicators to proactively address road risks and improve safety outcomes. IoT also empowers road safety by tapping into a wide range of previously unavailable data. Here are some benefits of this system [32]:

- Collection of traffic counts and real-time relay.
- Reduced wait time at signals and on highways.
- Proactive traffic management despite dynamic conditions.
- Striking the right balance between traffic density.
- Reduction in carbon monoxide emissions.
- Increasing road safety as this system leaves a minimal margin for error.

3.4 Smart Healthcare System

Smart healthcare is a collection of mobile devices, web-based applications, and digital technologies. Smart healthcare uses a combination of new-generation information technology solutions such as the IoT, cloud computing, AI, and big data to make healthcare more efficient and convenient [33]. Addressing modern healthcare care problems requires smart healthcare care management strategies. Hence, with integration, the government of India set a new benchmark through the Ayushman Bharat Digital Mission (ABDM), a basic principle supporting the country's integrated digital health infrastructure. Healthcare systems across India come under unprecedented pressure. Healthcare in India has long been provided in unconnected silos by individual physicians and small hospitals [34].

3.5 Smart Security

AI image analytics has revolutionized optical detection systems. Among other advantages, using AI in optical systems has increased their effectiveness and cost-effectiveness. A few of these benefits are:

- Reliable and consistent performance: AI-powered optical detection systems have several benefits, prominent among them the ability to operate continuously without downtime.
- Reduced operator workload and increased awareness: Optical detection systems with AI capabilities may properly and swiftly process massive amounts of data. They can quickly and accurately identify, monitor, and categorize threats by analyzing several real-time video feeds. Knowing how quickly threats may appear and change in today's fast-paced security environment is crucial.
- Increased precision: AI-enabled optical sensors improve the accuracy of drone detection systems by utilizing various customized machine-learning algorithms reinforced by feedback mechanisms. Additionally, items that are invisible to the human eyes, sometimes restricted to just one or two pixels in a video feed, can be detected by AI techniques.
- Ability to scale: Scalability is a key benefit of AI-powered optical systems. They may be used with other systems, such as RF and radar sensors, to form an all-encompassing defense network with several layers. AI-powered optical systems may give operators a more comprehensive image of the threat environment by integrating data from many sources, enabling them to make better judgments [35].

A few examples used in AI and IOT integrated OFS devices for smart security are SKYSIGHTTM and INSIGHTTM by OSL Technology (UK).

4 Benefits of Smart Infrastructure

i. **Economical**: It increases energy efficiency and effectively allocates resources by reducing costs.
ii. **Eco-friendly**: Reduce carbon footprint and focus on a greener future with pre-analysis information.
iii. **Well-being**: It provides easier access to essential services by reducing traffic jams and improving public safety, which can be possible by integrating infrastructure [35].

4.1 Overcoming Challenges in the Implementation of Smart Infrastructure

i. **Data Privacy and Security**: How a smart health program in Germany faced insufficient collection in data safety has been described [36]. Ensuring the privacy and security of the large amounts of data generated by smart infrastructure systems addressed critical challenges through robust cybersecurity measures and data governance policies.
ii. **Compatibility**: smooth and compatible smart integration is required for the full benefit of technologies.
iii. **Collaboration**: For successful smart infrastructure execution, key collaborators are required in the urban development team.
iv. **Efficiency**: cost-effective solution and the promotion of adaptable deployment.
v. **Data Management and Analysis**: Managing and analyzing vast amounts of data generated from IoT, AI, and optical fiber sensor integration is a significant challenge. Therefore, an advanced data center can efficiently process the data [37, 38].

5 Prospects

Integrated infrastructure and its modern solution have become urgent for cities to grow and evolve. Integrating modern technologies is laying the pathway for smart infrastructure, a revolutionized approach transforming design, build, and maintenance.

5.1 Empowering Urban Planning Using IoT

i. **Real-Time Data Capture**: Collection of valuable data on congestion patterns, energy consumption, air quality, and more can be done continuously by IoT-enabled embedded sensors throughout the city. Live monitoring helps city planners make informed decisions.
ii. **Improved Resource Allocation**: IoT analyzes the collected data for better planning and enhanced allocation for cost savings and higher efficiency.
iii. **Enhanced Citizen Engagement**: IoT-powered smart infrastructure can also facilitate citizen participation, allowing residents to provide feedback and participate in the decision-making process, community-building ownership, and collaboration [39].

5.2 *Harnessing Power of AI*

i. **Informed Maintenance**: It detects infrastructure failure, allowing active maintenance and reduced cost, enabling reliable service that extends the lifetime of critical components.
ii. **Automated Decision**: For live monitoring and computerized decisions, an AI algorithm is used, which improves efficiency and responsiveness.
iii. **Better Safety and Security**: Public safety can be modernized using facial recognition and video diagnostics to prevent incidents and enhance overall security [35].

5.3 *OFS in Monitoring*

i. **Structural Health Monitoring**: The incorporation of OFS sensors within the building and bridges enables live monitoring to detect faults for active maintenance, which reduces damages.
ii. **Environmental Monitoring**: Monitoring physical parameters using optical sensors provides accurate and comprehensive data collection for sustainability initiatives and urban planning decisions [11].
iii. **Asset Tracking**: OFS can track critical resources, such as water pipes or electrical cables, location, and status, for efficient maintenance and repair scheduling [40].
iv. **Nonlinear response**: The major disadvantage of the OFS sensor is a nonlinear (sigmoidal) response, which produces a limited useable linear region. Hence, to overcome such a problem, digital instrumentation and signal processing can be employed [41, 42].

6 Summary

The integration of optical fiber sensors (OFS), artificial intelligence (AI), and the Internet of Things (IoT) has enhanced productivity and quality of life, making it more sustainable and ecologically friendly. Integrating IoT and OFS has many benefits, such as reducing the number of sensors, which improves precision and reliability. IoT devices connect the physical devices to enable live monitoring in energy management, traffic control, network communication, and healthcare. The data set analyzed using AI improves operations and provides predictive information. This integration minimizes environmental impact, reduces downtime, increases life-useful benefits, and improves resource management. Innovative bridges with IoT and OFS devices continuously monitor structural integrity, strain, and potential faults. The power engineering community has been better equipped with these technologies to face challenges, enhance operational efficiency, and enable a sustainable and digitally enabled power ecosystem. These technologies are eco-friendly and economical and provide easier access to essential services. Monitoring physical parameters using optical sensors offers accurate and comprehensive data collection for sustainability initiatives and urban planning decisions. Despite certain challenges like data privacy, collaboration, and efficiency, the integration of AI and IoT with OFS devices has been of a great deal of interest in developing smart infrastructure.

References

1. S. Aleksic, A survey on optical technologies for iot, smart, industry, and smart infrastructures. J. Sens. Actuator Netw. **8**, 47 (2019)
2. K. Micko, P. Papcun, I. Zolotova, Review of IoT sensor systems used for monitoring the road infrastructure. Sens. **23**, 4469 (2023)
3. B. Astarloa, A. Kaakeh, M. Lombardi, J. Scalise, *The Future of Electricity: New Technologies Transforming the Grid Edge* (World Economic Forum, Geneva, Switzerland, 2017), pp. 1–32
4. M.L. Di Silvestre, S. Favuzza, E.R. Sanseverino, G. Zizzo, How decarbonization, digitalization, and decentralization are changing key power infrastructures. Renew. Sustain. Energy Rev. **93**, 483–498 (2018)
5. H. Amini, K. Boroojeni, S. Iyengar, F. Blaabjerg, P. Pardalos, A. Madni, A panorama of future interdependent networks: from intelligence infrastructures to smart cities. Sustain. Interdepend. Netw. Stud. Syst. Decis. Control, **145**, 1–10 (2018)
6. B. Gholamzadeh, H. Nabovati, Fiber optic sensors. World Acad. Sci. Eng. Technol. **42**, 297–307 (2008)
7. F. Matta, F. Bastianini, N. Galati, P. Casadei, A. Nanni, Distributed strain measurement in steel bridge with fiber optic sensors: validation through diagnostic load test. J. Perform. Const. Facil. **22**, 264–273 (2008)
8. K. Fritsch, Digital angular position sensor using wavelength division multiplexing. Proc. SPIE, Boston, MA, USA **1169**, 453–460 (1990)
9. T.D. Ponnimbaduge Perera, D.N.K Jayakody, S.K. Sharma, S. Chatzinotas, J. Li, Simultaneous wireless information and power transfer (swipt): recent advances and future challenges. IEEE Commun. Surv. Tutor. **20**, 264–302 (2018)
10. R. Shafin, L. Liu, V. Chandrasekhar, H. Chen, J. Reed, J.C. Zhang, Artificial intelligence-enabled cellular networks: a critical path to beyond-5G and 6G. IEEE Wirel. Commun. **27**, 212–217 (2020)
11. L.S.M. Alwis, K. Bremer, B. Roth, Fiber optic sensors embedded in textile-reinforced concrete for smart structural health monitoring: a review. Sens. **21**, 4948 (2021)
12. E. Esenogho et al., integrating artificial intelligence internet of things and 5g for next-generation smart grid: a survey of trends. Chall. Prospect. **10**, 4794–4831 (2022)
13. R. Alshawish, S. Alfagih, M. Musbah, Big data applications in smart cities. Proc. Int. Conf. Eng. & MIS, Agadir, Moroc. **22–24**, 1–7 (2016)
14. M. Batty, Big data, smart cities, and city planning. Dialogues Hum. Geogr. **3**, 274–279 (2013)
15. M. Batty, K. Axhausen, F. Giannotti, A. Pozdnoukhov, A. Bazzani, M. Wachowicz, G. Ouzounis, Y. Portugali, Smart cities of the future. Eur. Phys. J. Spec. Top. **214**, 481–518 (2012)
16. C. Balakrishna, Enabling technologies for smart city services and applications, in *Proceedings of the International Conference on Next Generation Mobile Applications, Services and Technologies* (Paris, France, 2012), pp. 223–227
17. W. Serrano, Digital systems in smart city and infrastructure: digital as a service. Smart Cities **1**, 134–154 (2018)

18. S.P. Mohanty, U. Choppali, E. Kougianos, Everything you wanted to know about smart cities: the internet of things is the backbone. IEEE Consum. Electron. Mag. **5**, 60–70 (2016)
19. T.H. Son, Z. Weedon, T. Yigitcanlar, T. Sanchez, J.M. Corchado, R. Mehmood, Algorithmic urban planning for smart and sustainable development: systematic review of the literature. Sustain. Cities Soc. **94**, 104562 (2023)
20. Hashem, Ibrahim, Chang, Victor, Anuar, Nor, S. Adewole, I. Yaqoob, G. Abdullah, E. Ahmed, H. Chiroma, Int. J. Inf. Manag. **36**, 748–758 (2016)
21. M. Ramakrishnan, Optical fiber sensors for smart composite materials structures, in *Optical Fiber Sensors: Advanced Techniques and Applications* (CRC Press, Boca Raton, FL, USA, 2015), pp. 491–520
22. V.M. Marukashan, On-line health monitoring of smart composite structures using fiber polarimetric sensors. Smart Mater. Struct. **8**, 544–548 (1999)
23. I. Machorro-Cano, J.O. Olmedo-Aguirre, G. Alor-Hernández, L. Rodríguez-Mazahua, L.N. Sánchez-Morales, N. Pérez-Castro, Cloud-based platforms for health monitoring: a review. Informatics **11**, 2 (2024)
24. J.J. Barriga, J. Sulca, J.L. León, A. Ulloa, D. Portero, R. Andrade, S.G. Yoo, Smart parking: a literature review from the technological perspective. Appl. Sci. **9**, 4569 (2019)
25. A.K. Sikder, A. Acar, H. Aksu, A.S. Uluagac, K. Akkaya, M. Conti, IoT-enabled smart lighting systems for smart cities, in *2018 IEEE 8th Annual Computing and Communication Workshop and Conference (CCWC)* (Las Vegas, NV, USA, 2018), pp. 639–645).
26. V.K. Quy, N.V. Hau, D.V. Anh, N.M. Quy, N.T. Ban, S. Lanza, G. Randazzo, A. Muzirafuti, IoT-enabled smart agriculture: architecture, applications, and challenges. Appl. Sci. **12**, 3396 (2022)
27. B. Darshan, M.K. Shashank, K. Srihari, K. Srinidhi, C.V. Reddy, Smart bridge. Int. Res. J. Eng. Technol. **7**(4) (2020)
28. I.S. Jha, Y.K. Sehgal, S. Sen, R. Kumar, *Smart Grid in Indian Power System NPSC* (2012)
29. I.S. Jha, S. Sen, R. Kumar, Smart grid development in India—a case study, in *2014 Eighteenth National Power Systems Conference (NPSC)* (Guwahati, India, 2014), pp. 1–6. https://doi.org/10.1109/NPSC.2014.7103866
30. C. Creß, Z. Bing, A.C. Knoll, Intelligent transportation systems using roadside infrastructure: a literature survey. IEEE Trans. Intell. Transp. Syst. **25**, 6309–6327 (2024)
31. Report on the Customized Driver Training Program titled, Defensive Driving and Driver Assistance through Artificial Intelligence, Conducted jointly by CSIR Central Road Research Institute & Ashok Leyland Institute for Driving Training & Research (as part of Project: iRASTE: Intelligent Solutions for Road Safety through Technology and Engineering
32. Ministry Of Urban Development, Traffic Management and Information Control Centre (TMICC) Operations Document, Government of India, Nirman Bhawan, New Delhi (2016)
33. C. Subbarao, et al., *Challenges for the Adoption of Smart Healthcare Strategies: An Indian perspective.* Publisher-British Academy of Management (2022). http://hdl.handle.net/2436/624809
34. R. Thakur, S. Sangar, B. Ram, M. Faizan, Quantifying the burden of out-of-pocket health expenditure in India. Public Health. **159**, 4–7. https://doi.org/10.1016/j.puhe.2018.02.017 pmid:29673888
35. N.H. Al-Ashwal, K.A.M. Al Soufy, M.E. Hamza, M.A. Swillam, Deep learning for optical sensor applications: a review. Sensors (Basel). **23**(14), 6486 (2023)
36. A.M. Selim, P.H.A. Yousef, M.R. Hagag, Smart infrastructure by (PPPs) within the concept of smart cities to achieve sustainable development. Int. J. Critical Infrastructures. **14**(2) (2018)
37. M.G. Melchiorre, R. Papa, M. Rijken, E. van Ginneken, A. Hujala, F. Barbabella, EHealth in integrated care programs for people with multimorbidity in Europe: insights from the ICARE4EU project. Health Policy **122**(1), 53–63 (2018)
38. Niti Ayog, *National Strategy for Artificial Intelligence* (Government of India, 2018)
39. M.E.E. Alahi, A. Sukkuea, F.W. Tina, A. Nag, W. Kurdthongmee, K. Suwannarat, S.C. Mukhopadhyay, Integration of IoT-enabled technologies and artificial intelligence (AI) for smart city scenario: recent advancements and future trends. Sens. **23**, 5206 (2023)

40. A. Gharaibeh et al., Smart cities: a survey on data management, security, and enabling technologies. IEEE Commun. Surv. & Tutor. **19**, 2456–2501 (2017)
41. Z.M. Çınar, A. Abdussalam Nuhu, Q. Zeeshan, O. Korhan, M. Asmael, B. Safaei, Machine learning in predictive maintenance towards sustainable smart manufacturing in industry 4.0. Sustain. **12**, 8211 (2020)
42. M.N. Taib, R. Narayanaswamy, Novel technique to improve optical fiber sensor's response for industrial applications, in *IEEE Colloquium on Advances in Sensors* (London, UK, 1995), pp. 13/1–13/6

Case Studies: AI-Enhanced Optical Fiber Sensors in Healthcare and Environmental Monitoring

Medha Bhushan, Rekha Bhardwaj, Gayathri Rajeshwari Nathavajhula, and Parmita Sarigala

Abstract In recent years the need for more robust environmental and healthcare technology has grown and optical fiber sensors enhanced with artificial technology might be the answer. Sensors that make use of the unique properties of optical fiber and AI trained on studies and previous case studies can provide greater comfort to users and improve quality of life. These sensors can be used in the collection of data related to healthcare and environmental monitoring amongst others. Wearable sensors such as wristbands and watches have improved flexibility owing to the development in optical fibers and found their place in healthcare, able to track vital signals of the body. Using such advanced technology in monitoring the climate, checking the pollution levels in the air and soil will be tremendously beneficial. By making use of a different fiber optic sensors, integrated with AI & ML, it enhances the system's ability to analyze spectral shifts and minimizing errors, enabling accurate and robust health and environmental monitoring. In this we hope to bring light on the various applications of AI enhanced optical fibers in health and environment and the technology behind it.

Keywords Fiber optic sensors AI enhanced optical fibres · Healthcare monitoring · Environmental monitoring

1 Introduction and Overview

An important technological advancement that offers capabilities much beyond those of conventional fiber optics is the combination of artificial intelligence (AI) and optical fiber sensors. This chapter examines the possibilities of AI-enhanced optical

M. Bhushan (✉) · G. R. Nathavajhula · P. Sarigala
Department of Basic Sciences, G. Narayanamma Institute of Technology and Science (For Women), Shaikpet, Hyderabad, Telangana, India
e-mail: drmedha.bhushan@gnits.ac.in

R. Bhardwaj
Department of Applied Sciences, Bharati Vidyapeeth College of Engineering, New Delhi, India

V. Bhardwaj et al. (eds.), *Optical Fiber Sensors and AI*, Progress in Optical Science and Photonics 34, https://doi.org/10.1007/978-981-96-6158-9_9

fiber sensors, emphasizing their benefits over traditional systems and their new uses in environmental and medical monitoring [1].

Real-time data analysis, subtle change detection, and even problem prediction is all possible with AI-enhanced optical fiber sensors. The chapter will begin with the basic characteristics that set AI-based fiber sensors apart from traditional ones, including their capacity for prediction, daily real-time data processing, and enhanced sensitivity [2].

By offering previously unattainable levels of precision for determining the conditions these patients fall under, we will examine how they are being utilized in healthcare to revolutionize patient monitoring and diagnostics. Similar to environmental monitoring, recording and managing change triggered by artificial intelligence itself has become possible with AI-trained optical fiber sensors for use in hazardous environments, offering a new degree of accuracy [3].

These case studies aim to essentially reveal the wide range of AI-enabled optical fiber sensors, highlighting their consequences and potential applications in these core fields.

2 Advantages of AI-Enhanced Optical Fiber Sensors Over Traditional Fiber Sensors

2.1 Understanding Optical Fiber Sensors

Optical fiber sensors are like the nervous system of many modern technologies. They use light to sense and measure changes in their surroundings, such as temperature, pressure, or stress. These sensors have been around for a while, helping us monitor and maintain everything from bridges to medical equipment [1]. While traditional fiber sensors have been incredibly useful, adding artificial intelligence (AI) to these sensors takes their capabilities to a whole new level in the following ways.

2.2 Making Sense of Data, Instantly

One of the biggest advantages of AI-enhanced optical fiber sensors is their ability to make sense of data in real-time. Traditional sensors are great at collecting data, but they often rely on humans or external software to interpret that data. This process can be slow and sometimes misses important details. With AI, the sensor doesn't just gather data; it analyzes it on the spot. This means patterns and potential issues are identified immediately, leading to quicker and more informed decisions [2].

2.3 *Catching the Small Stuff*

AI doesn't just help with speed; it also makes sensors much more sensitive. Traditional sensors do a good job of picking up changes, but they might overlook tiny variations that could be significant. AI-enhanced sensors can detect even the smallest shifts, which is especially important in critical situations like monitoring a building's structural integrity or diagnosing a patient. This heightened precision means fewer mistakes and better outcomes [3].

2.4 *Seeing Problems Before They Happen*

Imagine if your car could tell you a week in advance that your brakes are about to fail. That's the kind of predictive power AI brings to optical fiber sensors. Traditional sensors can alert you when something is wrong, but AI takes it a step further by predicting potential problems before they occur. This is possible because AI can analyze past data and recognize patterns that signal trouble ahead. The result? You can fix issues before they become costly disasters, saving time and money [2].

What if your vehicle could alert you a week before your brakes are likely to fail? AI gives optical fiber sensors that kind of predictive power. While conventional sensors can notify you when anything is amiss, artificial intelligence goes one step further by anticipating possible issues before they arise. This is made feasible by AI's ability to examine historical data and identify trends that point to potential problems. The outcome?

You can save time and money by resolving problems before they become expensive catastrophes [1].

2.5 *Learning and Adapting on the Fly*

One of the most impressive things about AI is its ability to learn and adapt. Traditional sensors operate within fixed parameters you set them up, and they do their job. But AI-enhanced sensors can adjust themselves based on new information. This adaptability is crucial in environments that are constantly changing. Whether it's adjusting to shifts in weather or evolving industrial processes, AI allows these sensors to stay accurate and effective without needing constant human input [4].

2.6 Less Work, More Results

AI-enhanced sensors are also great at reducing the need for human involvement. Traditional sensors often require regular maintenance, calibration, and data interpretation—all tasks that can be time-consuming and prone to error. With AI, many of these tasks are automated. The sensors can calibrate themselves, analyze data without outside help, and even fix minor issues on their own. This automation not only saves time but also reduces the chance of mistakes, leading to more reliable systems [5].

2.7 Keeping Data Safe and Secure

In a world where data security is increasingly important, AI adds an extra layer of protection to optical fiber sensors. In sensitive applications, like in defense or finance, the accuracy and privacy of data are crucial. AI can help detect and prevent security breaches, ensuring that the data collected remains safe. Moreover, AI can verify that the data hasn't been tampered with, providing greater confidence in the information these sensors provide [6].

2.8 A Smart Investment

While AI-enhanced optical fiber sensors might cost more upfront, they often prove to be a smart investment in the long run. The ability to predict maintenance needs, reduce downtime, and minimize human intervention all contribute to lowering overall costs. Plus, the improved accuracy and reliability of these sensors help avoid expensive mistakes and system failures. Over time, the savings and efficiencies gained can outweigh the initial expense [7].

In summary, AI-enhanced optical fiber sensors are a step forward in technology, offering significant improvements over traditional fiber sensors. By making sensors smarter, more sensitive, and capable of learning, AI is transforming how we monitor and manage systems across a wide range of industries. As AI technology continues to advance, these sensors will likely become even more powerful tools, making our lives easier, safer, and more efficient [2].

3 Health Monitoring Using AI Enhanced Optic Fiber Sensors

Optical Fiber Sensors and Healthcare

An overview of optical fiber sensors and their advantages have been discussed in the previous section. Owing to their numerous benefits, lightweight, compactness, chemical stability and resistance to electromagnetic field to name a few, optical fibers have great potential in the field of healthcare. An early diagnosis is the best prevention to diseases that can become life threatening without timely medical intervention [8].

No matter how great the capabilities of healthcare professionals are, without proper equipment to diagnose patients, it is impossible for them to perform to the best of their abilities. Optical fiber sensors have numerous uses in the world of medicine and have gained great acclaim. Wearable sensors are compact, easy to use, cheap and widely available to the public. These perform various roles, including health monitoring, remote monitoring and home rehabilitation [9, 17]. The most widely used optical fiber sensors are Fiber Bragg Grating sensors whose applications will be majorly discussed.

Artificial Intelligence in Optical Sensors

Movement and physiological data are gathered via wearable sensors, enabling the patient's health state to be tracked. A range of clinical-related sensors are used to assess a patient's critical health indicators, including blood pressure, blood glucose, body temperature, oxygen saturation, and pulse rate. The data values obtained from these sensors are then transmitted to the cloud using Internet of Things (IoT)-based systems.

Depending on their needs, the channel may choose to make the data they send public or private. Following that, the data is analysed using machine learning (ML) techniques. The two main categories of machine learning algorithms are supervised and unsupervised learning algorithms. Supervised learning methods are employed when the objective is to predict the values for the training dataset.

Fibre Bragg Grating Sensors

Distributed Bragg reflectors in a brief optical fiber segment make up the Fiber Bragg Grating (FBG) sensor, which transmits all other wavelengths of light while reflecting a specific one. This function can be accomplished by employing ultraviolet light to produce a wavelengthspecific dielectric mirror and permanently periodic refractive index fluctuation along the fiber core's propagation axis. When an FBG is stretched, its grating period changes, changing the Bragg wavelength [10].

$$\lambda = 2n\wedge$$

where n is the effective refractive index, λ is the grating period, and λ is the Brag-wavelength.n and Λ will change in tandem with changes in the observed physical quantity (e.g., temperature, stress, etc.) used for the fiber grating, which will result in λ drift and, in turn, by detecting the drift of λ. It is also possible to get information about the actual quantity being measured. The primary emphasis of research on Bragg fiber grating sensors is quasidistributed temperature and stress sensing. The formula for a shift brought on by variations in temperature and stress is $\Delta\lambda = \alpha\Delta T$ $\Delta\lambda = \varepsilon S$, where S is the grating's strain sensitivity, $\Delta\lambda$ is the wavelength change, α is the fiber's temperature coefficient, ΔT is the temperature change, and ε is the fiber's strain.

Applications in Healthcare

There are both invasive and non-invasive uses for optical fiber sensors. They are easily put into human cavities due to their small size and non-galvanic nature. Furthermore, they don't conduct electricity. Additionally, electromagnetic disturbances from electrical equipment (e.g., MRI, wireless communication, or electrosurgery) do not impact optical fiber sensors [10].

Case Study 1: Femur Fracture and Strain Monitoring

A femur fracture is a potentially fatal injury that is typically brought on by a high energy impact and is linked to significant local soft-tissue damage in addition to other more severe axial injuries. The current standard of care for this injury is early surgical stabilisation, which enables early mobilisation and rehabilitation.

To provide stability at the fracture site, an intramedullary device or plate is typically used.

It may take a few weeks or months to recover from a tibia fracture. Determining the degree of fracture recovery and the best time for the patient to return to normal load-bearing activities is challenging. CT scans expose the patient to dangerous radiation, even though they can give an approximate indication of fracture recovery. In order to monitor the strain and stiffness in bone and orthopaedic implants, alternative sensors that use optical fibres with Fibre Bragg Gratings (FBGs) are utilised [11].

Compared to standard resistive sensors, which detect the average strain between two specified spots or insertion sites within a ligament, FBG sensors are advantageous since they offer strain over the entire tissue via multiplexing techniques. Furthermore, electrical circuitry may harm resistive sensors, while FBGs are best suited for clinical applications because of their low conductivity. They are more sensitive and resistant to electromagnetic fields, lighter, smaller, and simpler to embed.

Case Study 2: Respiratory Monitoring

One crucial aspect of human health is respiration rate. When a person breathes, their chest and belly typically inflate during inhalation and droop during exhalation, creating a breathing action. Chest and abdomen displacement caused by respiratory activity results in λ shift, which is then sent to FBG, according to the principle of FBG respiratory monitoring equipment based on chest and abdomen movement.

Because the package material is wrapped around the chest and belly, it will distort as a result of the movement of the chest and abdomen during breathing. At the conclusion of breath and exhale, respectively, the sensor records the maximum and minimum deformation. The two deformations' spacing and duration can be computed to determine the frequency of breathing and cycle [12].

Case Study 3: Joint Angle Monitoring

The most widely used physical rehabilitation therapy application is the wearable joint angle monitoring gadget. Currently, the FBG-based joint angle monitoring device is applicable to every joint in the body. One major factor contributing to the extensive use of physical rehabilitation therapy is the long-term, ongoing monitoring of the joints. In conventional medical diagnosis, the measurement is typically done with a handheld goniometer, which takes a long time to record data and has low precision.

This conundrum has been significantly altered by the introduction of joint monitoring devices based on optical fiber sensors, which are increasingly being used in medical monitoring. A flexible material encases the FBG sensor, which is secured to the joint with an elastic belt. The flexible material will stretch, distort, and transmit to the FBG when the joint flexes and expands [12].

In addition to the applications listed above, wearable optical fiber sensors based on FBG can be used to monitor blood pressure, temperature, and heart rate.

Wearable Sensors

The market currently offers wearable sensors in the form of smart watches, gloves, patches, tattoos, facemasks, wrist bands, clothes, and glucometers that measure blood pressure, oxygen level, temperature, glucose level, body fluids, pulse and heart rate, electrolytes, medications, hormones, and metabolites. They are designed with exceptional wearing comfort in addition to strong sensing capabilities. Wearable sensor manufacturing now heavily relies on the choice and packaging of flexible materials and components [9, 12, 13].

Case Study 4: Tremor Severity in Parkinson's Disease

The Unified Parkinson's Disease Rating Scale (UPDRS), which is comparable to how neurologists grade scores in real clinical practice, was predicted using machine learning algorithms in this Seoul University study. A wrist-watch-style wearable device with a gyroscope and accelerometer was used in this study to evaluate the tremor signals of 85 Parkinson's disease (PD) patients. The acceleration, angular velocity, displacement, and angle signals were utilized for analysis after the displacement and angle signals were computed from the recorded acceleration and angular velocity.

The primary discovery is a machine-learning-based computerized scoring system for the severity of Parkinsonian tremor that is incredibly accurate. When comparing the five UPDRS classes to the conventional UPDRS that neurologists measure, an RMSE of 0.410 was found. Our method's accuracy is 85.5%, and this error is the smallest for automatic scoring of full range UPDRS in resting tremor [14].

Case Study 5: Monitoring Brain Injury

At Imperial College London, an AI-enabled optical sensor device was created to evaluate traumatic brain injury (TBI) biomarkers. It demonstrated encouraging results in animal tests. Throughout treatment, TBIs must be closely observed. Because of this, intracranial probes are used in neurocritical care settings to track biomarkers, or important indicators of the course of injury, such as oxygen and pressure in the brain.

Only one biomarker can be measured at a time by some of these probes. Others are capable of monitoring many indicators, but doing so necessitates inserting multiple tubes into the brain, which increases the danger of infections or additional tissue damage.

After traumatic brain damage, Imperial researchers have now created a patient monitoring system to track several biomarkers. The gadget combines machine learning algorithms that use historical data to forecast biomarker concentrations based on collected data in real-time with the capability of simultaneously monitoring four biomarkers.

Here, the dissolved oxygen (DO), temperature, pH, and glucose levels in brain artificial cerebrospinal fluid (aCSF) were all dynamically monitored at the same time using a multiplexed optical fiber sensor coupled with a programmed AI prediction platform. These four biomarkers were selected due to their significance in brain physiology.

Hospitals may be able to monitor TBI more efficiently if the device is refined and confirmed to be useful in humans [15].

4 Environmental Monitoring Using AI Enhanced Optic Fiber Sensors

Environmental monitoring plays a crucial role in safeguarding our planet and ensuring sustainable development. With the increasing complexity of environmental challenges and the need for precise, real-time data, the integration of advanced technologies has become essential. One such technological advancement is the combination of Artificial Intelligence (AI) with fiber optic sensors (FOS). This integration has revolutionized the way we monitor and manage various environmental factors, offering unprecedented levels of accuracy and efficiency.

Optical fiber sensors operate on the principle of light propagation within a flexible glass or plastic fiber. By detecting changes in light characteristics such as intensity, phase, or wavelength, these sensors can provide valuable data on physical, chemical, and biological phenomena. When integrated with advanced artificial intelligence (AI) algorithms, the capabilities of these sensors are significantly enhanced, enabling real-time data analysis and predictive insights that were previously unattainable [16].

Types of Fiber Optic Sensors Using AI

This report aims to explore the impact of integrating AI with fiber optic sensors across various domains of environmental monitoring. We will examine a few areas, particularly; Air Quality Monitoring, Water Quality Monitoring, Soil/ Land Monitoring, Radiation Monitoring, and Natural Disaster Monitoring and Prediction [16].

As we delve into the various applications and benefits of AI-enhanced fiber optic sensors, we will uncover the potential of these technologies in addressing some of the most pressing environmental challenges of our time.

Case Study 1: Air Quality Monitoring

Fiber Optic Based Gaseous Sensors

Photonic Crystal Fibers (PCFs) have transformed the world of optical sensing. It is a sophisticated optical fiber with precisely arranged microscopic air holes along its length. These air holes are integral, allowing PCFs to manipulate light in ways that traditional fibers cannot. This capability makes PCFs highly suitable for advanced applications, including environmental monitoring and industrial safety.

One standout application of this technology is the hollow-core PCF-based Mach–Zehnder interferometer (MZI) sensor, designed specifically to detect CO^2 gas [16]. This sensor, encased in gas-permeable membranes, not only responds quickly within 70 s but also adjusts to environmental conditions like temperature fluctuations, making it highly adaptable for real-world situations.

By combining AI with these advanced sensors, we can push their capabilities even further. AI steps in to process and analyze the sensor's data in real-time, sifting through the noise to deliver more accurate readings. It can even dynamically recalibrate the sensor based on changes in the environment, ensuring peak performance under varying conditions.

Moreover, AI can monitor the health of the sensor itself, predicting when maintenance or recalibration might be needed before any issues arise. In safety- critical situations, AI can detect anomalies in sensor data, flagging unexpected events that could indicate potential hazards. This allows the sensors more reliable, faster, and capable of tackling the complex challenges of modern industrial and environmental monitoring.

Monitoring of Volatile Organic Compounds (VOCs)

Volatile Organic Compounds (VOCs) are integral to natural processes and interactions between plants and animals [17]. However, VOCs resulting from human activities pose significant health risks, including respiratory issues and damage to vital organs. As environmental and health concerns rise, monitoring VOC concentrations in air and water becomes increasingly vital. Traditional detection methods such as mass spectrometry, gas chromatography, and high-performance liquid chromatography; offer precision but are often costly, bulky, and time-consuming. MEMS and MOS transistor-based sensors provide a more compact and cost-effective solutions;

however, these sensors operate at high temperatures (200–400 °C), which diminishes their low-power advantages.

Recent innovations have led to the development of advanced VOC sensors. One notable design incorporates photonic integrated technology with a chemically selective polydimethylsiloxane (PDMS) polymer cladding on an arrayed waveguide interference platform. This sensor, operational at room temperature, exhibits excellent selectivity for VOCs such as toluene, chlorobenzene, and hexane, achieving a maximum spectral shift of 22.8 nm for chlorobenzene at 1566.7 nm. It features a sensitivity of 234.8 pm/% and a detection limit of 0.24% m/m. AI algorithms enhance this sensor's capabilities by processing and analyzing complex data to reveal patterns and correlations that improve detection accuracy and reliability.

Additionally, silicon nitride (SiN) waveguide-based mid-IR sensors, manufactured using low-pressure chemical vapor deposition (LPCVD), offer enhanced sensitivity due to a broader mid-IR transparent zone and stronger evanescent wave. These sensors demonstrated a five-fold increase in sensitivity compared to traditional silicon waveguides, effectively monitoring VOCs such as acetone, ethanol, and isoprene. AI's real-time data analysis capabilities further bolster these sensors, enabling immediate detection of abnormal VOC levels and facilitating prompt responses to potential environmental or industrial hazards.

Innovative optical fiber VOC sensors include long-period grating (LPG) sensors with calixarene coatings, which are highly sensitive to aromatic compounds like benzene and toluene. Fiber Bragg grating (FBG) sensors, coated with PDMS, detect VOCs through Bragg wavelength shifts caused by PDMS swelling. Moreover, LPG sensors coated with silk fibroin have been developed for methanol detection, achieving a high sensitivity of 0.22 nm/mbar. AI integration in these sensors, particularly through predictive maintenance, allows for forecasting when sensors might need upkeep or repair based on historical data and usage patterns, thereby ensuring consistent performance and minimizing downtime [18].

Case Study 2: Water Quality Monitoring

Fiber Optic Sensors for Water Quality Monitoring

Optical sensors can also be designed to detect variations occurring in aquatic environments. This novel low-cost device, equipped with dual photodiode detectors and a multi-wavelength light source, excels in measuring water transparency and identifying potential pollution. Integrating AI into this sensor's operation could enhance its performance by enabling machine learning algorithms to analyze light transmission patterns, thus improving the accuracy of pollution event predictions and the overall efficiency of environmental monitoring. Today's fiber grating sensors provide unparalleled detail in temperature-depth profiles. With AI enhancements, these sensors could evolve from measuring conditions for predicting environmental changes, making our understanding of ocean dynamics even more precise.

A recent innovation involves a hollow square-core fiber (HSCF) sensor, designed to monitor ethanol–water evaporation with remarkable sensitivity. Integrating AI into this system could refine its detection capabilities, allowing for more accurate predictions and adjustments based on real-time data.

Fiber Bragg grating (FBG) sensors, known for their sensitivity to sea water temperature and pressure, have seen significant advancements. AI can further elevate these sensors by enabling real-time data analysis and detecting anomalies, thereby providing more insightful environmental monitoring. Lastly, a cutting-edge embedded waveguide-based fluorescent immunosensor can identify up to 32 pollutants, including microcystin-LR in lake water. AI integration can amplify its efficiency, allowing for quicker and more accurate detection of multiple pollutants, thus enhancing our ability to safeguard water quality [19, 20].

Water pH Monitoring Using FOS

Potentiometric sensors have emerged as a crucial tool for measuring pH levels. These sensors operate using sensitive and reference electrodes to provide precise pH measurements based on Nernstian principles. Among various sensor types, those utilizing RuO_2 as the sensing element stand out for their high sensitivity and stability. These advanced sensors cover a broad pH range from 2 to 13 and deliver a sensitivity of 58 mV/pH at 23 °C [21]. However, they face a challenge: their response time increases at lower temperatures, taking 8–10 min at 9 °C compared to just 1–2 s at 23 °C [20].

The integration of AI technology could revolutionize these sensors, making them even more effective. Machine learning algorithms can analyze data from these potentiometric sensors to enhance their accuracy. AI can refine calibration methods, adjust for temperature fluctuations, and predict sensor drift, ensuring consistent performance across diverse environmental conditions. This fusion of AI with traditional sensor technology promises a leap forward in precise and reliable aquatic monitoring [21].

Case Study 3: Land Monitoring

Advancements in Soil Moisture Monitoring via FOS

Traditional soil moisture sensors include resistive, capacitive, dielectric, and densiometric types. However, these sensors often encounter limitations when monitoring large areas due to the need for numerous sensors and extensive cabling.

Fiber optic sensors offer a promising alternative for soil moisture monitoring. Their ability to provide continuous, real-time measurements with minimal disturbance makes them particularly well-suited for monitoring small-scale soil processes. FOS can cover extensive areas thanks to their multiplexing capabilities, which allow for comprehensive monitoring with fewer sensors [22, 23].

Several advancements have been made in fiber optic sensors for soil moisture measurement. For example, the use of optical fibers to measure soil water content by observing changes in light transmission. While this study demonstrated that light transmission decreases as soil water content increases, it did not include sensor calibration.

Similarly, the use of Fiber Bragg Grating (FBG) technology has also been explored. Laylor et al. developed a sensor incorporating two FBGs—one for temperature and one for both temperature and humidity [22]. This sensor, housed in a hydrophilic medium, effectively detected soil moisture but was limited to a range of 2 to 4% moisture. Machine learning models analyze historical and current data to forecast future soil moisture levels, allowing proactive adjustments. With the integration of AI, it improves irrigation scheduling, detects anomalies, integrates diverse data sources, and adapts over time, offering precise and actionable insights for farmers [22, 24].

Case Study 4: Radiation Monitoring

Radiation Monitoring in Nuclear Reactors

In the intricate world of nuclear reactors and power plants, ensuring safety and efficiency is paramount. Radiation monitoring in these high-stakes environments is critical to maintaining operational integrity and protecting human health. Fiber optic sensors have emerged as a revolutionary technology in this domain, offering precise and reliable radiation detection. When these sensors are enhanced with artificial intelligence, their capabilities are significantly amplified. AI integration transforms these fiber optic sensors into intelligent systems capable of real-time analysis and predictive maintenance.

The fusion of fiber optic technology and AI leads to a sophisticated monitoring system where data from the sensors is not merely collected but actively analyzed and interpreted. This synergy allows for the detection of minute changes in radiation levels, which could indicate potential issues before they escalate into serious problems. The AI algorithms process vast amounts of data with remarkable speed and accuracy, providing actionable insights that support timely decision-making. This proactive approach to radiation monitoring helps mitigates risks and ensures the continuous safe operation of nuclear reactors and plants.

Furthermore, AI-enhanced fiber optic sensors contribute to optimizing the maintenance schedules and operational strategies of these facilities. By analyzing historical data and identifying patterns, Artificial Intelligence can predict equipment failures or radiation leaks, enabling preemptive actions that can save both time and resources. The result is a more resilient and efficient nuclear power infrastructure, where the integration of cutting-edge technology ensures a safer environment for both workers and the surrounding communities [8, 25].

Radiation Monitoring in Space Vehicles

Space exploration presents unique challenges, and one of the critical aspects of ensuring mission success is monitoring radiation levels within space vehicles and during space missions. The harsh environment of space exposes astronauts and equipment to high levels of radiation, making effective monitoring essential for the safety of crew members and the integrity of space missions. Fiber optic sensors, combined with the power of artificial intelligence, provide a groundbreaking solution to this challenge.

In the vacuum of space, where traditional radiation detection methods may fall short, fiber optic sensors offer a robust and reliable alternative. High-purity germanium detectors are known for their high-resolution capability in detecting gamma radiation. These detectors can be coupled with fiber optic technology to improve their performance in the space environment [26]. AI-enhanced systems can analyze the data from these detectors to provide accurate radiation assessments. Similarly, Silicon detectors, including silicon photomultipliers (SiPMs), are sensitive to various types of radiation and can be used in space missions. These sensors, when enhanced with AI, deliver unprecedented accuracy and responsiveness in monitoring radiation levels. The AI algorithms embedded in the sensors analyze real-time data to detect fluctuations in radiation, allowing for immediate responses to potential hazards. This integration ensures that astronauts are shielded from harmful radiation levels and that mission equipment operates within safe parameters.

The combination of fiber optic sensors and AI also revolutionizes the way data is utilized in space missions. AI-driven analytics not only provide real-time radiation monitoring but also predict radiation exposure trends based on historical data and mission parameters. This forward-looking approach enables mission planners to devise strategies for radiation protection and equipment maintenance, optimizing the overall success and safety of space missions.

Case Study 5: Natural Disaster Monitoring and Prediction

Earthquake Monitoring Using FOS

Monitoring undersea earthquakes and studying the Earth's interior has long been a challenge due to the difficulties in deploying and maintaining seafloor geophysical apparatus. Recent advancements in fiber-optic sensor technology offer a revolutionary approach by utilizing existing underwater communications cables. These sensors analyze changes in the polarization of optical telecommunication channels to detect seismic and water waves. A notable example is the use of a 10,000-km cable stretching from Los Angeles to Valparaiso, Chile, which has successfully monitored moderate-to-large earthquakes and identified potential tsunami events. This method is highly scalable, transforming global underwater cables into continuous, real-time observatories without the need for specialized equipment or dedicated fibers [27].

Artificial Intelligence significantly enhances fiber-optic seismic monitoring systems in multiple ways. Machine learning algorithms are pivotal in processing the vast amounts of data generated by these sensors. By analyzing historical seismic data, AI models can identify patterns that signal seismic events. These algorithms, particularly supervised learning models, can detect anomalies and differentiate between genuine seismic signals and background noise. AI's ability to process data in real time ensures that even subtle changes are promptly identified, enabling rapid detection of potential seismic events and timely alerts.

Predictive modeling is another area where AI excels. By recognizing complex patterns and correlations in seismic data, AI can refine predictive models. Machine learning models integrate various data sources, such as temperature, pressure, and seismic activity, to enhance forecasting accuracy and reduce false positives. As new data is continuously fed into these models, AI systems update their predictions, improving the accuracy of earthquake forecasts and their potential impacts.

Furthermore, AI enhances the sensitivity and accuracy of fiber-optic seismic monitoring sensors. Advanced algorithms can filter out noise from the seismic data, isolating meaningful signals and improving the sensors' sensitivity. This allows for the detection of smaller seismic events that might otherwise be missed. AI's adaptive systems can also adjust to new seismic patterns, ensuring that the monitoring system remains accurate and reliable over time.

Seismic monitoring systems are categorized into acoustic and radiation-based methods. Acoustic systems focus on the registration of infrasound and ultrasound to detect seismic activity. Distributed Acoustic Sensing (DAS) technology, one of the most developed methods, uses fiber-optic cables installed specifically for research or existing unused fibers within optical networks. To enhance sensitivity, optical fibers are equipped with Fiber Bragg Gratings (FBGs), which reflect optical beams at locations where the fiber's properties change due to local vibrations. This method allows for the detection of seismic activity over large distances.

Radiation-based methods, on the other hand, involve detecting radon gas concentrations. High radon levels are indicative of potential earthquakes. Fiber-optic sensors can detect alpha particle emissions from radon gas, with systems such as Fiber Bragg Grating arrays and Fabry–Perot resonators increasing the sensitivity of these detections. This method can predict earthquakes one to two weeks before significant seismic events [16].

Landslide Monitoring Using FOS

Understanding the dynamics of landslides is crucial for developing effective early warning systems. To achieve this, innovative monitoring techniques are required. One such technique involves using optical fiber sensors embedded within a large-scale physical model of a slope to measure strain induced by landslides. This approach utilizes distributed optical fiber sensing technology, specifically optical frequency domain reflectometry, to provide high-resolution strain measurements along the slope.

In a groundbreaking experiment conducted within a 6-by-2 m reinforced-concrete structure simulating a slope, an optical fiber sensor cable was meticulously positioned at the predetermined failure surface. This setup allowed for real-time monitoring of strain evolution with centimeter-level spatial precision. The fiber sensor cable, approximately 35 m in length, was arranged in four primary straight spans descending longitudinally along the slope, each spaced at 0.25, 0.75, 1.25, and 1.75 m from the left lateral wall. Aluminum profiles connected to the lateral walls were mechanically secured to these spans, creating a well-defined measurement system for stress and strain analysis [16].

To validate and compare the fiber-optic measurements, conventional sensors were also employed to measure pore water pressure and soil moisture at various depths and locations along the slope. This comprehensive approach ensured robust data collection and provided a comparative framework for analyzing the optical fiber's performance.

The sensor system demonstrated unprecedented clarity in monitoring the strain field's evolution during the landslide event. The real-time data acquisition revealed detailed insights into the landslide's development stages, from the initial strain changes to the eventual collapse. The precision of the optical fiber measurements allowed for the identification of early warning indicators well before the collapse occurred. This capability is critical for developing advanced early warning systems that can provide timely alerts and potentially mitigate the impact of landslides. The experiment's design also included provisions to account for temperature fluctuations, which could affect the sensor's accuracy.

Artificial Intelligence significantly augments the functionality of fiber-optic landslide monitoring systems by introducing advanced data analysis and predictive capabilities. In particular, machine learning algorithms, including supervised and unsupervised learning methods, are employed to analyze the complex and voluminous data produced by optical fiber sensors.

AI techniques, such as neural networks, are capable of recognizing intricate patterns within the strain data that may not be evident through conventional analysis. These algorithms can detect subtle changes in strain that precede a landslide, improving the system's sensitivity and predictive accuracy. For instance, a deep learning model can be trained on historical strain data associated with past landslides, allowing it to identify similar patterns in real-time data and provide early warnings.

Moreover, AI-powered systems can implement anomaly detection algorithms to isolate significant deviations from normal strain patterns. These deviations could indicate early warning signs of slope failure, such as sudden increases in strain or shifts in strain distribution. The integration of AI also facilitates adaptive learning, where the system continually refines its predictive models based on new data. This dynamic learning process allows the system to improve its forecasts over time, adapting to changing environmental conditions and enhancing its overall effectiveness. In addition, AI algorithms can be used to process data from multiple sensors simultaneously, creating a comprehensive picture of the landslide's progression. This multi-sensor fusion enables more accurate and timely detection of potential failures, integrating information from various points along the slope to deliver a cohesive analysis.

Volcanic Eruptions Using FOS

Despite significant advancements in volcanology, the intricacies of volcanic eruptions remain elusive due to the complex internal structures of volcanoes, which can obscure and distort observable signals. Traditional methods struggle to detect subtle precursor activities and the precise mechanisms leading to eruptions, leaving critical gaps in our understanding and predictive capabilities [27].

Recent innovations have introduced optical fibers coupled with Distributed Acoustic Sensing (DAS) as transformative tools for volcanic monitoring. This approach leverages the sensitivity of optical fibers to detect and analyze acoustic signals associated with volcanic activity [28]. DAS technology enables the remote identification of volcanic occurrences and provides detailed imaging of subsurface volcanic features that were previously hidden from view.

In a study, researchers employed DAS technology to monitor volcanic activity by analyzing strain signals related to eruptions. By utilizing a 2D template matching technique, they could correlate detected signals with expected wave arrivals, accurately pinpointing the origins of volcanic explosions. This method uncovered vital details about volcanic structures, including the detection of a scoria layer with spatially variable thickness exhibiting complex grain interactions. This advanced capability offers a more nuanced understanding of volcanic dynamics and improves our ability to detect minor eruptions and fluid movements within the volcano.

The integration of AI further refines volcanic monitoring capabilities. AI algorithms excel in mastering signal processing, enabling the extraction of meaningful patterns from complex datasets. Through convolutional neural networks and other machine learning techniques, AI can discern subtle variations in strain signals that might indicate imminent volcanic activity. This capability extends to predictive modeling, where AI analyzes historical and real-time data to forecast potential eruptions and provide early warnings, thereby enhancing preparedness and risk management.

By identifying deviations from established patterns, AI can alert scientists to unusual volcanic behaviors that may precede significant events. Furthermore, AI integrates data from various sources, including optical fibers, seismic sensors, and satellite imagery, creating a comprehensive and accurate picture of volcanic activity.

The integration of optical fiber technology, DAS, and AI represents a significant leap forward in volcanic monitoring. This advanced approach not only improves our understanding of volcanic processes but also enhances our ability to predict and manage volcanic hazards. As we continue to develop and refine these technologies, we pave the way for more effective surveillance and risk mitigation strategies, ultimately safeguarding communities and advancing our knowledge of the Earth's dynamic systems.

Flood Monitoring Using FOS

In flood-prone regions, early and accurate water level monitoring is vital for disaster prevention. Traditional methods struggle with the complexity and variety of flood scenarios, highlighting the need for advanced, cost-effective sensor systems. Optical fiber sensors have emerged as a transformative tool in water level monitoring. These sensors offer significant advantages, including immunity to electromagnetic interference, lightweight construction, and safety in explosive environments. They come in various forms, such as: continuous, discrete, intrusive, and non-intrusive; catering to different monitoring needs [29].

Recent advancements in sensor technology involve using plastic optical fibers (POFs), which can be easily shaped without breaking. This flexibility simplifies the manufacturing process and enhances the sensor's durability compared to traditional glass fibers, which are prone to breakage and sensitivity issues.

A notable innovation is the wireless mesh network using POF sensors for flood monitoring. This system employs wireless motes with MICA2DOT™ units to capture and transmit data from the sensors to a central station. The POF sensors, equipped with a V-shaped probe, detect changes in water levels by observing variations in optical signals [29].

AI enhances this system by continuously analyzing data to forecast flooding events, thus improving the accuracy and efficiency of flood monitoring. This integration not only enhances measurement accuracy but also enables advanced data analysis for early flood detection [30].

5 Conclusions and Outlook

There are several uses for optical fiber in a variety of industries, including healthcare and telecommunication. We can realize the full potential of optical fiber by augmenting it with the newest, cutting-edge technologies, such as machine learning and artificial intelligence.

In this research, many characteristics and special benefits that optical fibers provide are examined, and some of their noteworthy uses in environmental monitoring and healthcare are mentioned. The combination of these innovative technologies has shown encouraging progress in a number of important areas. The use of artificial intelligence (AI) in optical fiber anomaly detection has given rise to new examples ith generative models that promise improved precision and dependability in the detection of anomalies in fiber optic vibrations. These systems can also analyze intricate spatiotemporal data patterns in real-time.

An advancement in technology, AI enhanced optical fiber sensors provide notable advantages over conventional fiber sensors.

AI is revolutionizing the way we monitor and operate systems in a variety of industries by making sensors smarter, more sensitive, and able to learn. These sensors will probably develop into even more potent instruments as AI technology advances, improving our quality of life and making it safer, easier, and more effective. AI-enhanced optical fibers are expanding the realm of monitoring and diagnostic capabilities in the healthcare industry. These sensors provide unmatched sensitivity and precision for anything from tracking joint angles and respiration patterns to assessing biomarkers in brain injury, opening the door to more efficient and individualized medical care.

Our capacity to monitor and assess air quality has greatly increased because to AI-enhanced fiber optic sensors, such as those that use Photonic Crystal Fibers and sophisticated VOC detection techniques. In addition to providing accurate, real-time data, these sensors also take advantage of AI's powers in anomaly identification and predictive maintenance, which guarantees increased responsiveness and dependability in environmental and industrial contexts. The field has undergone a revolution thanks to the integration of AI with fiber optic sensors for water quality, which has improved the precision and effectiveness of pH monitoring and pollution identification. A proactive approach to environmental preservation is made possible by AI's ability to analyze complicated data and forecast changes in water quality, allowing for quicker and more efficient solutions to problems involving water contamination. AI and fiber optic sensors have demonstrated significant promise in monitoring natural disasters like floods, landslides, earthquakes, and volcanic eruptions. AI can enhance early warning systems and prediction models by evaluating data from optical fiber sensors, providing sophisticated capabilities for risk management and catastrophe preparedness. By overcoming the difficulties associated with traditional monitoring, this technique improves flood management and forecast, which helps with disaster response and mitigation.

The continuous developments in this area imply that AIenhanced optical fiber sensors will be essential in influencing innovation and technology in the future, propelling advances in a variety of sectors and raising our standard of living.

References

1. R. Kumar, K. Gupta, R. Singh, *Application of Artificial Intelligence in the Healthcare Sector: A Critical Analysis*, pp. 387–412 (2024). https://doi.org/10.1007/978-981-97-7815-7_23
2. Y. Zhou, Y. Zhang, Q. Yu, L. Ren, Q. Liu, Y. Zhao, Application of machine learning in optical fiber sensors. Meas. **228**, 114391 (2024). https://doi.org/10.1016/j.measurement.2024.114391
3. X. Tang, X. Li, Y. Ding, M. Song, Y. Bu, The pace of artificial intelligence innovations: Speed, talent, and trial-and-error. J. Informetr. **14**(4), 101094 (2020). https://doi.org/10.1016/j.joi.2020.101094
4. T. Sun et al., Artificial intelligence meets flexible sensors: emerging smart flexible sensing systems driven by machine learning and artificial synapses. Nano-Micro Lett. **16**(1), 14 (2023). https://doi.org/10.1007/s40820-023-01235-x
5. Z. Zhang, L. Wang, C. Lee, Recent advances in artificial intelligence sensors. Adv. Sens. Res. **2** (2023). https://doi.org/10.1002/adsr.202200072
6. S.K. Devineni, AI in data privacy and security. Int. J. Artif. Intell. Mach. Learn. **3**, 35–49 (2024)
7. A. Venketeswaran et al., Recent advances in machine learning for fiber optic sensor applications. Adv. Intell. Syst. **4**, 2100067 (2021). https://doi.org/10.1002/aisy.202100067
8. J. Castrellon-Uribe, Opt. Fiber Sens.: Overv. (2012). https://doi.org/10.5772/28529
9. F. Khoshmanesh, P. Thurgood, E. Pirogova, S. Nahavandi, S. Baratchi, Wearable sensors: at the frontier of personalised health monitoring, smart prosthetics and assistive technologies. Biosens. Bioelectron. **176**, 112946 (2021). https://doi.org/10.1016/j.bios.2020.112946
10. A.D. Shakir, Optical fiber sensors for biomedical applications of optical fibers, fiber sensors, impact of light and healthcare industry trends. J. Curr. Med. Res. Opin. Receiv. **06**(11), 1849–1855 (2023)
11. A. Najafzadeh et al., Application of fibre bragg grating sensors in strain monitoring and fracture recovery of human femur bone. Bioeng. **7**, 98 (2020). https://doi.org/10.3390/bioengineering7030098
12. J.S. Kim et al., Wearable hand module and real-time tracking algorithms for measuring finger joint angles of different hand sizes with high accuracy using FBG strain sensor. Sens. **20**(7) (2020). https://doi.org/10.3390/s20071921
13. S. Junaid et al., Recent advances in artificial intelligence and wearable sensors in healthcare delivery. Appl. Sci. **12**, 10271 (2022). https://doi.org/10.3390/app122010271
14. H. Jeon et al., Automatic classification of tremor severity in Parkinson's disease using a wearable device. Sens. **17**(9) (2017) https://doi.org/10.3390/s17092067
15. A.G. Leal-Junior, C.A.R. Diaz, L.M. Avellar, M.J. Pontes, C. Marques, A. Frizera, Polymer optical fiber sensors in healthcare applications: a comprehensive review. Sens. **19**(14) (2019). https://doi.org/10.3390/s19143156

16. M.A. Butt, G.S. Voronkov, E.P. Grakhova, R.V Kutluyarov, N.L. Kazanskiy, S.N. Khonina, Environmental monitoring: a comprehensive review on optical waveguide and fiber-based sensors. Biosens. **12**(11) (2022). https://doi.org/10.3390/bios12111038
17. A. Russo, S. Pollastri, M. Ruocco, M.M. Monti, F. Loreto, Volatile organic compounds in the interaction between plants and beneficial microorganisms. J. Plant Interact. **17**(1), 840–852 (2022). https://doi.org/10.1080/17429145.2022.2107243
18. L. Manjakkal et al., Connected sensors, innovative sensor deployment, and intelligent data analysis for online water quality monitoring. IEEE Internet Things J. **8**(18), 13805–13824 (2021). https://doi.org/10.1109/JIOT.2021.3081772
19. D. Lyu, Q. Huang, X. Wu, Y. Nie, M. Yang, Optical fiber sensors for water and air quality monitoring: a review. Opt. Eng. **63** (2023). https://doi.org/10.1117/1.OE.63.3.031004
20. A. Omar, M.Z. Mat Jafri, Development of optical fiber sensor for water quality measurement. **1017** (2008). https://doi.org/10.1063/1.2940669
21. T.H. Nguyen et al., Fluorescence based fibre optic pH sensor for the pH 10–13 range suitable for corrosion monitoring in concrete structures. Sens. Actuators B Chem. **191**, 498–507 (2014). https://doi.org/10.1016/j.snb.2013.09.072
22. M. Leone, (INVITED)Advances in fiber optic sensors for soil moisture monitoring: a review. Results Opt. **7**, 100213 (2022). https://doi.org/10.1016/j.rio.2022.100213
23. L. Chen, C. Xia, Z. Zhao, H. Fu, Y. Chen, AI-driven sensing technology: review. Sens. **24**(10) (2024). https://doi.org/10.3390/s24102958
24. C.V.N. Bhaskar, S. Pal, P.K. Pattnaik, Recent advancements in fiber bragg gratings based temperature and strain measurement. Results Opt. **5**, 100130 (2021). https://doi.org/10.1016/j.rio.2021.100130
25. J. Wu et al., Distributed fiber sensors with high spatial resolution in extreme radiation environments in nuclear reactor cores. J. Light. Technol. **39**(14), 4873–4883 (2021). https://doi.org/10.1109/JLT.2021.3075630
26. A. Rovera, A. Tancau, N. Boetti, M.D.L. Dalla Vedova, P. Maggiore, D. Janner, Fiber optic sensors for harsh and high radiation environments in aerospace applications. Sens. **23**(5) (2023). https://doi.org/10.3390/s23052512
27. B. Anuradha, C. Abinaya, M. Bharathi, A. Janani, A. Khan, IoT based natural disaster monitoring and prediction analysis for hills area using LSTM network, in *2022 8th International Conference on Advanced Computing and Communication Systems (ICACCS)* (2022), pp. 1908–1913. https://doi.org/10.1109/ICACCS54159.2022.9785121
28. Z. He, Q. Liu, Optical fiber distributed acoustic sensors: a review. J. Light. Technol. **39**(12), 3671–3686 (2021). https://doi.org/10.1109/JLT.2021.3059771
29. S. Subramaniam, V.R. Gannapathy, S. Subramonian, A. Hamidon, Flood level indicator and risk warning system for remote location monitoring using flood observatory system. WSEAS Trans. Syst. Control **5**, 153–163 (2010)
30. P. Macheso, The role of optical fiber sensor in internet of things: applications and challenges: review (2023). https://doi.org/10.13140/RG.2.2.21755.49445

Cybersecurity and Ethical Considerations in AI-Augmented Optical Fiber Sensing

Umashankar Samal, Shatakshi Singh, Garima Nain, and Amit Das

Abstract Optical Fiber Sensing (OFS) technology is termed a highly sensitive and reliable method for monitoring physical conditions, such as temperature, humidity, pressure, and strain, across myriad and often difficult environments. Comprehended for its durability and low power consumption, OFS has evolved as an indispensable component in applications ranging from structural health monitoring to environmental and industrial sensing. Artificial Intelligence (AI) integration into OFS systems has recently elevated its capabilities. It facilitates improved data processing, decision-making, anomaly detection, knowledge extraction, and predictive insights, qualifying for real-time and adaptive responses in crucial situations. Nevertheless, the OFS and AI augmentation are prompting cybersecurity and ethical challenges. As AI-integrated OFS systems are implanted across sensitive sectors that range from healthcare to national security and critical infrastructure, the risks associated with data security and privacy, unauthorized access, and bias in AI models turn out to be increasingly important. This study thoroughly explores the cybersecurity vulnerabilities and ethical considerations associated with AI-driven OFS technology. This work highlights the significance of robust cybersecurity protocols and ethical standards to unlock AI's full potential in OFS applications responsibly.

Keywords Artificial intelligence · Cybersecurity · Ethical considerations · Optical fiber · Optical fiber sensing

---

Author to whom any correspondence should be addressed.

U. Samal (✉)
Department of Mathematics, GLA University, Mathura, Uttar Pradesh, India
e-mail: umashankar.samal249@gmail.com

S. Singh · A. Das
Department of Electronics and Communication Engineering, GLA University, Mathura, Uttar Pradesh, India

G. Nain
Department of Information Technology, Atal Bihari Vajpayee Indian Institute of Information Technology and Management, Gwalior, Madhya Pradesh, India

V. Bhardwaj et al. (eds.), *Optical Fiber Sensors and AI*, Progress in Optical Science and Photonics 34, https://doi.org/10.1007/978-981-96-6158-9_10

1 Introduction

Sensors are paramount building blocks in numerous Internet of Things (IoT) applications and cyber-physical systems [1]. These devices detect and respond to various physical and environmental inputs, such as temperature, pressure, light, motion, or moisture [2]. As an enabler, these devices habilitate the physical information (in terms of signals or data) of the environment into the digital record [3]. These sensors enable systems to make informed decisions or perform specific actions by analyzing the data collected in digital form.

Revolutionalizing and constituting myriad application domains, for instance, in industrial applications, these sensors are integrated to monitor machinery for performance, helping to prevent breakdowns and maintain efficiency [4]. In healthcare, sensors like heart rate and oxygen monitors provide real-time data on patient vitals, supporting early diagnosis and continuous health monitoring [5, 6]. Environmental sensors are used to track weather patterns, air quality, and pollution levels, which helps scientists and governments in planning for climate resilience [7, 8]. In daily life, sensors play a key role in smart devices like phones, where accelerometers detect device orientation, and in home automation systems, where motion sensors activate lighting or security alerts [9, 10]. Businesses can gain insights for focused marketing campaigns by analyzing social media data to understand behavior and interests. Motion detection and heat sensors are used in security applications to monitor activities and movements, improving safety in specified regions. In addition, the sensor-driven insights are valuable for enhancing structural health and safety while optimizing industrial and civil construction operations. The adaptability and preciseness of sensors make them paramount across a broad spectrum of fields, improving efficiency, safety, and convenience.

The amount of data produced, communicated, and exchanged through these sensors has increased to a huge scale globally [11]. Thus, obtaining real-time decision-making, enhanced security, and efficacy to enhance the planning processes in numerous sectors becomes necessary. The effort toward digital transformation further intensified the initiatives to automate and fuse data across numerous sectors, such as healthcare, manufacturing, agriculture, transportation, environmental monitoring, and industry. They facilitate expedited, data-informed decisions while preserving time, effort, and resources. It allows informed choices at remarkable speeds.

In this regard, optical fiber sensing (OFS) technology is evolving as a paramount instrument to meet the growing demand for accuracy, speed, and robustness [12]. OFS is a highly progressive technology using optical fibers (small glass or plastic strands). It transmits light through the optical fibers to detect changes in physical parameters such as temperature, strain, pressure, and vibration. It is recognized for its sensitivity, durability, and capability to monitor large areas with minimal power consumption, making it beneficial for long-term applications in challenging environments. Unlike traditional sensors, OFS systems utilize the behavior of light within the fiber to detect external disturbances.

This characteristic makes them ideal for continuous, real-time monitoring in numerous application domains or fields. OFS systems are resistant to electromagnetic interference, durable in extreme circumstances, and competent in providing distributed sensing over large areas. These benefits make them admiringly preferable in comparison to traditional sensors.

Some prime application sectors of the OFS are illustrated in Fig. 1 and discussed in detail in the following section. It showcases its versatility in healthcare, environmental monitoring, transportation, smart cities, and infrastructure management.

Structural Health Monitoring: The OFS technology has its roots in the civil engineering sub-domain to estimate the structural integrity of monuments, bridges, buildings, tunnels, underpasses, and pipelines. It facilitates early detection of stress, strain, and temperature interpretations from the abovementioned civil structures, which are used to reveal the potential structural failures from these sensory inputs. Herein,

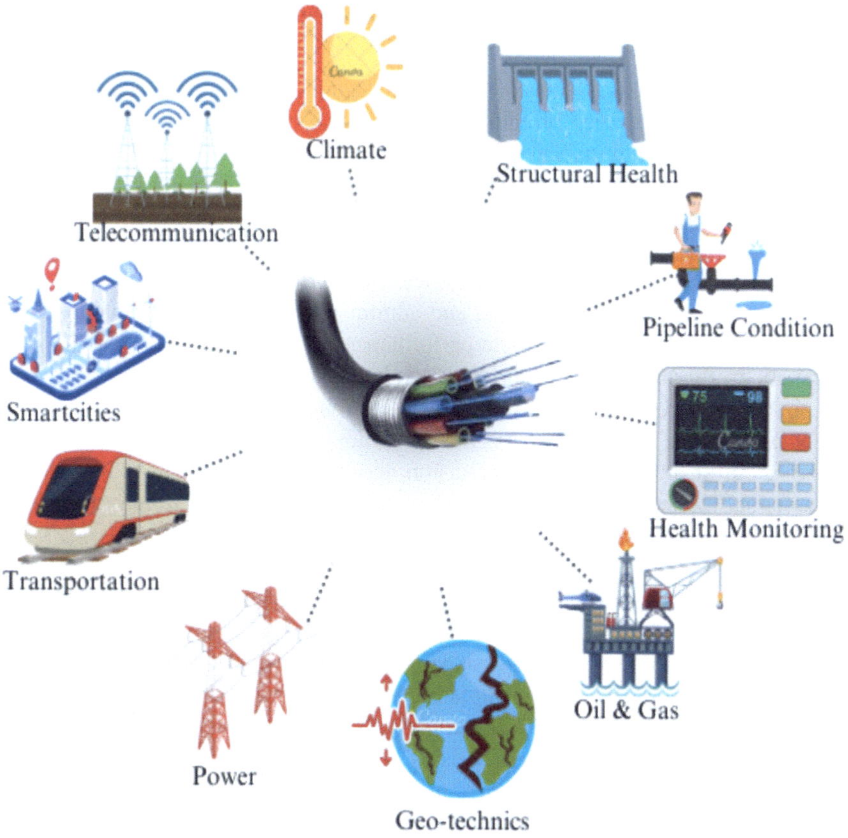

Fig. 1 Applications of OFS

the OFS plays a critical role in this application for preventing accidents, detecting cracks, and structural maintenance while lowering maintenance costs by delivering actionable data that allows proactive repairs.

Medical Diagnostics: OFS is likewise embraced and engaged in the medical and healthcare domain, especially in the context of minimally invasive diagnostics. Due to its biocompatibility and flexibility, it is particularly suitable for endoscopic diagnosis procedures and for the monitoring of physiological parameters such as heart rate, oxygen levels, blood pressure, and respiratory function. It contributes to the precise and early detection of diseases and thus improves patient health monitoring in real-time monitoring, being a paramount manifestation in diagnostics.

Environmental Sensing: Another widely used application of the OFS technology is environmental monitoring due to its sensitivity, robustness, and adaptability under miscellaneous and critical situations. These recognize crucial environmental modifications such as changes in water quality, air pollution levels, humidity levels, and temperature variations. These sensors are particularly useful in challenging or remote circumstances, including undersea ecosystems, volcanic eruption-prone regions, and extreme climates, where typical electrical sensors fail or degrade over time.

Industrial Applications: OFS is indispensable for upholding operational safety and efficacy in various industries, providing high-precision monitoring, maintenance, and inspection solutions. For instance, it is extensively utilized in the oil and gas industry to minimize the risk of environmental hazards and costly interruptions by detecting leaks, recognizing structural vulnerabilities, and monitoring the industry. Moreover, it assures the integrity of data transmission within the telecommunications sector, which is essential for the secure and seamless flow of information in industries such as finance, e-governance, healthcare, and administration. The automotive industry uses OFS to assess real-time vehicle performance in advanced safety systems. It is extremely useful in autonomous driverless vehicles and smart parking systems that rely on the quality of the sensory input collected from the surroundings. At the same time, the aerospace industry employs this technology to monitor tension and temperature in aircraft designs. Its high precision and reliability render it indispensable and essential for sectors relying on continuous, uninterrupted operations and quick response to potential hazards.

On the other hand, Artificial Intelligence (AI) is designated as a transformative technology that mimics or stimulates human intelligence to analyze data, learn from patterns, and make autonomous decisions [13]. Using machine learning and deep learning methods, AI analyzes extensive input data from various sources at tremendous paces and accuracy levels, acclimating and improving with each iteration. This capability makes AI a vital mechanism across myriad industries where efficiency, precision, and quick response are necessary [14, 15]. In the context of Industry 5.0, the fifth industrial revolution, AI is pivotal in cultivating a collaborative environment between humans, machines, and robots. The objective is to develop manufacturing and operational systems that are autonomous, smarter, and more adaptable while focused on sustainability. AI is used in Industry 5.0 for automation and advanced data analysis, quality inspection, predictive maintenance, product design, and intelligent decision-making.

When AI technology is applied to OFS, it dramatically enhances its capabilities. In the OFS systems, AI assists in analyzing complex data signals from fiber sensors, identifying anomalies, and developing predictive insights. This integration converts OFS into an advanced diagnostic and monitoring system capable of real-time, automated, and data-driven decision-making. AI-enabled OFS can detect subtle environmental changes, structural stress patterns, machine predictive maintenance, and smart inspection systems that could remain unnoticed when performed manually. It additionally improves reliability, automation, safety, and responsiveness in applications varying from infrastructure health monitoring to environmental sensing, as mentioned in detail in Fig. 1.

As the OFS increasingly merges with AI mechanisms, the handling of cybersecurity and ethical concerns becomes necessary since AI-enabled data processing and decision-making create new vulnerabilities. This transpires as the AI-integrated OFS systems handle tremendous amounts of sensitive data within different industrial and operational applications for training and inference. Unauthorized access, data breaches, and possible tampering threaten the application systems' integrity and reliability. This results in false readings or malicious manipulation with significant ramifications for safety and privacy.

Ethically, AI-augmented OFS systems introduce concerns regarding data ownership, permission, and proper use of personal and environmental data. Therefore, the data collection and processing procedure must adhere to data privacy compliance while yielding to relevant rules. In this context, it is crucial to establish clear data security regulations, ensure openness in data consumption, and embed robust cyberthreat-related defenses. This is important for building trust and accountability in deploying AI-integrated OFS solutions.

In light of the above discussion, this chapter explores the cybersecurity vulnerabilities and ethical implications of integrating AI into OFS applications. As AI enhances the capabilities of OFS across diverse applications, it also brings about new risks and ethical considerations that must be carefully managed and resolved for efficacious AI-integrated OFS systems. The scope of this chapter includes an examination of potential security threats, like data breaches and system manipulation, along with ethical challenges regarding data privacy, ownership, and responsible data usage.

2 Cybersecurity Threats in AI-Augmented OFS

As mentioned before, incorporating AI into OFS will result in a substantial improvement in data processing, monitoring, decision-making, and analytics. It delivers lofty sensitivity and precision across miscellaneous applications, from structural health and industrial monitoring to environmental sensing. Nevertheless, the inclusion of AI in OFS technology presents cybersecurity vulnerabilities that must be addressed to preclude unauthorized access, secure data integrity, and support system reliability. This section illustrates the immediate cybersecurity threats associated with AI-augmented OFS systems, including data vulnerabilities and possible mitigation approaches.

Data Vulnerabilities in AI-Augmented OFS: Access to huge data is crucial for AI-integrated OFS systems to function effectively, especially for deep learning models that require huge training data. This facilitates real-time data collection, processing, training, analysis, and decision-making. Hence, it mandates a constant data flow, which exposes it to several cyberthreats. This exposure is crucial, particularly in critical sections of industries that deal with privacy and security-critical data, such as critical infrastructure, industrial regulations and compliances, national security, and patients' personal healthcare information. The key risks within this context include data tampering, unauthorized access, and interception, significantly compromising system reliability and data integrity.

Data Interception: The data traversed from OFS sensors to centralized processing systems or the computing device deployed with the AI algorithms can be precluded by attackers. Hence, it creates huge opportunities for interception to unauthorized access and data manipulation due to continuous data flow in AI-assisted OFS systems. Hackers can impede the transmission of this data, removing sensitive knowledge or altering it for malicious intentions. As an illustrative example, the OFS systems in the structural health monitoring data might reveal vulnerabilities. In another example regarding national security applications, precluded data can grant unauthorized parties access to surveillance information, posing risks to secure installations. If this information is accessed or manipulated, it could compromise the safety and integrity of the monitored structures. Thus, from these instances, it is interpreted that protecting the privacy and security of OFS data is essential.

Unauthorized Access: AI-driven OFS designs typically implicate comprehensive data storage to retain real-time knowledge about structural health and environmental conditions. In the absence of stringent authentication and access controls, it leads to the prominent risk of unauthorized access. This can additionally lead to potential data breaches, misuse, or even malicious manipulation of stored sensory data. Thus, unauthorized access to OFS data can seriously affect sectors that employ AI-driven OFS. An attacker gaining unauthorized access to environmental monitoring data might alter climate reports, misrepresenting the real environmental conditions. Similarly, inappropriate access to patients' medical information obtained through OFS could endanger their confidentiality and also manipulate it to hinder efficient diagnosis. Thus, to resolve the above issues, it is required to implant multi-layered authentication mechanisms and access control mechanisms to protect the data transported by OFS systems integrated with AI.

Data and AI Models Tampering: Another factor is data tampering in AI-augmented OFS systems by the attackers, which can yield high risk. Due to data tampering, the sensor data is modified, manipulated, or disrupted and fed for the training of the AI model. This results in data that ought to skew and mislead interpretations, projections, and predictions. Therefore, OFS systems with data tempering usher to incorrect readings or missed detection of crucial events, especially in critical security and safety applications in industries such as oil, gas, and chemical leaks. Attackers not only alter the OFS-based collected data or insert false signals but additionally access and alter the trained AI model, thus compelling it to misinterpret sensory outputs. This is achieved by adjusting the AI model's parameters or weights, indirectly modifying its relationships between input–output and detection thresholds. In such scenarios, false alarms or, more importantly, missed detections are extremely common, endangering safety and accuracy, which are crucial and of utmost priority.

These data susceptibilities highlight the importance and vigorous requirement of robust cybersecurity frameworks in AI-enhanced OFS systems. Therefore, protecting and safeguarding the data collection, transmission, storage, integrity, and prediction pipelines ensures the security of information generated by OFS systems. Moreover, it also ensures the safety and reliability of the applications employing the OFS.

3 Mitigation Strategies for Data Vulnerabilities in AI-Augmented OFS

Multifarious strategies can be employed to mitigate data vulnerabilities and enhance the overall cybersecurity of AI-driven operational framework systems. Below are some recommended measures for mitigation:

Encryption of Data Transmission: Encryption is key for defending against data interception and securing information flow between OFS sensors, AI algorithms, and centralized systems. By encrypting the data in the storage and transmission stages, AI-driven OFS systems can prevent unauthorized users from accessing sensitive information. The advanced encryption protocols (end-to-end encryption) being developed nowadays are particularly effective in securing the communication media between OFS sensors and data processing units at the computing platforms. It helps ensure data interception that remains inaccessible without the inaccessible decryption keys.

Multi-Factor Authentication (MFA) and Access Control Mechanisms: Implementing robust authentication measures, such as multi-factor authentication (MFA), guarantees that only authorized users can access the OFS-generated data. In addition, the access control mechanisms restrict unauthorized access to data pipelines and reduce the risk of data tampering or breaches. For example, role-based access control (RBAC) methods can limit access to collected data depending on their roles, whereas MFA embeds extra security by mandating multiple layers of identification.

Data Integrity Checks: Constant and regular data integrity checks can help identify data tampering, which is achieved by validating the consistency and authenticity of sensory data collected by OFS. These include techniques such as hash functions and checksums, which are employed to ensure that any unauthorized alterations and modifications have been detected early. In accordance with the alterations, the management and addressing of the problem can be achieved. In addition to this, blockchain technology is touted as a means of recording data transactions in a secure, immutable ledger. It delivers trustless exchanges by eradicating the need for intermediaries through cryptographic technology and makes the AI-integrated OFS systems more secure, safe, and protected.

Anomaly Detection Systems: AI technology has been used in detail to analyze the data for anomaly detection. In this contrast, AI-enabled OFS systems can embed AI-based anomaly detection systems to recognize unusual patterns or behaviors in data effectively. Therein, it enables early detection of potential cyberthreats before utilizing this data for analytics. It is accomplished by monitoring the data for drifts and deviations from their expected density function or behavior. To this end, these systems are capable of detecting and reacting to security concerns on time. For example, in an anomaly detection system, irregularities in environmental monitoring data can be spotted. After identifying the possible threats, a security alert could be activated, providing a detailed investigation to decide whether the irregularities belong to data tampering or unauthorized access.

Secure AI Training Protocols and AI models: The robustness of AI models used in OFS systems relies significantly on the data quality and privacy in model training. On-device training mechanisms could maintain the privacy and security of the training data. AI-integrated OFS systems can mitigate the possibilities of data poisoning or adversarial attacks by embracing secure training protocols such as local training mechanisms named federated learning and data sanitization. Federated learning allows models to be trained locally on the decentralized data and devices without sharing the raw and sensitive data among the existing training devices. It minimizes the risk of unauthorized access and can be achieved in cross-silo or cross-device, horizontal or vertical federated learning.

In light of the above discussion, a comprehensive procedure is necessary to address the data vulnerabilities in AI-augmented OFS. This approach must include data encryption, robust authentication methods, integrity checks, and privacy-enhanced AI training processes. These strategies are essential for protecting the integrity and confidentiality of OFS data and ensuring reliability and resilience against potential cybersecurity threats. This is key to the OFS-based secure monitoring of vital infrastructure, healthcare, industrial, and environmental conditions. Implementing these cybersecurity standards is crucial for defending public safety and ensuring optimal system performance in a globally interconnected world.

4 Ethical Considerations

Integrating AI into OFS systems requires ethical consideration, especially in data-sensitive applications involving personal data, data transparency, and consent. In this section, a brief discussion of the ethical concerns involved in AI-driven OFS systems is incorporated. It possesses problems such as privacy risks or concerns, fairness and biases in the AI algorithms, the need for transparent and accountable decision-making, and surveillance and consent issues.

Privacy Concerns: The privacy concerns in the AI-enabled OFS systems are an outcome of their ability to gather and analyze significant quantities of sensitive data. It is evident that some of the data collected from the OFS may contain private and sensitive data. This is especially true in healthcare, environmental tracking, and security surveillance applications that capture detailed information about physical environments, persons, locations, and individuals. Thus, private or personally identifiable information (PII) is at risk when OFS is combined with AI algorithms, while deriving insights may inadvertently expose private information. It leads to potential misuse or unauthorized surveillance. These privacy concerns are made even stronger by the fact that a lot of data needs to be kept for a long time for machine learning to train and test. Furthermore, transmitting and storing such data boosts the risk of data breaches or cyberattacks. To address these concerns, we require robust encryption, data anonymization techniques, and strict compliance with privacy regulations, such as GDPR, to ensure the ethical and secure deployment of AI-enabled OFS technologies. This ensures and establishes the protection of individuals' privacy.

Bias and Fairness in AI Algorithms: The AI algorithms are susceptible to bias, and thus it is a possible ethical issue in OFS systems using AI algorithms. The AI model inadvertently incorporates biases that lead to skewed or unfair predictions during the training on the domain-specific training data. An example is employing AI-enabled health monitoring of buildings; the training data containing bias toward a certain geographical region will lead to biased predictions and inaccurate results. Identically, biasedness can be seen in algorithms that can affect certain groups or communities causing differences in outcomes or the distribution of resources. Handling these issues requires meticulous validation of training data and algorithms to detect and correct potential biases, promoting fairness and inclusivity. This will make AI-driven OFS applications more fair and open to everyone.

Transparency and Accountability: Transparency and explainability in AI-based decision-making are essential for reliability and accountability. In AI-augmented OFS systems, determinations and interpretations of the sensory input can be complex. It is important since often the predictions made by the AI models are termed as "black-box" as they are beyond human interpretations and predictions. Thus, in security and safety–critical applications, a lack of explainability and transparency can lead to big ethical queries. Ensuring transparency involves visualizing the AI models and their explainability using the final or intermediate predictions. It brings the trust of the user to the predictions made by the AI model. Accountability is crucial to identifying responsible parties for AI-generated errors or misinterpretations that lead to adverse effects. Consequently, AI-augmented OFS systems must gather the trust of users by prioritizing transparency, explainability, and accountability.

Surveillance and Consent: One of the more sensitive ethical issues surrounding AI-powered sensing systems is the possibility of unauthorized surveillance and the necessity of obtaining informed consent. AI-augmented OFS systems deployed in public or semi-private spaces, such as urban cities, workplaces, or hospitals, could inadvertently monitor individuals without their knowledge or consent. This introduces ethical concerns regarding individuals' rights to privacy and sovereignty. This is so because they might be unaware of the extent to which their personal data, actions, or behaviors are recorded, observed, and analyzed. To address these concerns, providing clear communication about the purpose, scope, and limitations of surveillance in AI-OFS applications is essential. Additionally, mechanisms should be in place to obtain informed consent from affected individuals. AI-augmented OFS systems should prioritize privacy-respecting designs, such as limiting data collection to non-identifiable information or focusing solely on environmental metrics instead of personal data.

5 Case Studies

Integrating AI into OFS systems has unfurled unique capabilities across sectors but also brings up cybersecurity risks and ethical concerns. In the following section, we have provided a detailed description of the two case studies. The first is regarding AI-enhanced OFS in national security and the other is in healthcare, concentrating on the ethical and cybersecurity matters that come with these technologies.

Case Study 1: AI-Augmented OFS in Healthcare

AI-augmented OFS systems are used in healthcare and medicine for progressive diagnostics and continuous patient monitoring. As mentioned earlier in Sect. 1, OFS systems attached to wearables and handheld devices are used widely in the healthcare sector to measure blood pressure, oxygen levels, heart rate, and respiratory functions. They can additionally be connected in an IoT network by sending the collected data to medical professionals for analytics. Along with the healthcare professionals and doctors this data additionally fed to the AI models for early diagnosis of the health problems. This acts as proactive remediation and improves patient health monitoring.

Cybersecurity Issues: Despite offering valuable insights, these AI-enabled systems in healthcare are vulnerable to cybersecurity threats. Unauthorized access to sensitive and personal patient data is a prime concern, as this data often contains PII and sensitive medical components. The alteration of this data may lead to incorrect diagnoses or false negatives that additionally impact the contamination and spread of diseases such as COVID-19. The AI models can also become targets for adversarial attacks, where a malicious user exploits the input data to produce false health issues. The implementation of complex encryption protocols and secure AI training methods is necessary to mitigate these risks. In addition, strict access controls should be implemented to protect patient information and system integrity.

Ethical Concerns: Using AI-enabled OFS in patient monitoring raises ethical queries about privacy and consent, as the patients might not be fully aware of the procedure of the data collection and AI-based interpretation of their health. There is also the issue of transparency and explainability when achieving AI-based decision-making. These procedures might be difficult for patients and health professionals to comprehend. This brings a higher degree of trust issues since the diagnosis or prediction cannot be readily explained. These ethical concerns can be addressed by providing appropriate information to the patients regarding the type of data collected, proper consent forms, and explainable AI methods by healthcare institutions.

Case Study 2: AI-Augmented OFS in National Security

In the national security domain, AI-powered OFS systems are being embraced for applications such as border surveillance, critical infrastructure monitoring, equipment health monitoring, environmental forecasting, and defense systems. For instance, fiber-optic sensors installed along the border premises of a country can detect vibrations, motions, actions, proximity, and sounds. AI mechanisms then analyze the collected data to interpret the regular, human, animal, and environmental factors like wind. This encourages security forces to quickly detect the threats and respond accurately to these unusual scenarios. In this context, it maintains the security and safety of the defense personnel, equipment, weapons, and borders, thereby enhancing national security.

Cybersecurity Issues: As the susceptivity of nations' security and the data related to it is well known, AI-augmented OFS systems are the key targets for cyberattacks. Data poisoning by malicious attackers allows them to manipulate sensor readings, access organized information, or alter predictions. Adversarial attacks on AI methods comprehend false positives or missed detections. This undermines the data corresponding to the nation's security efforts. Defending these systems from attackers and cyberattacks requires a combination of encrypted data transmission, advanced intrusion detection systems, and regular assessments of AI model robustness. This can indeed prevent the system from being exploited and ensure its reliability.

Ethical Concerns: Using AI for surveillance in nations' security contexts presents significant ethical challenges, especially concerning privacy and civil liberties. Conducting surveillance without informed consent raises important queries about the tradeoff between security and individual rights. For example, citizens may be monitored without their knowledge or permission. Furthermore, AI models employed for nations' security applications may hold biases, leading to unfair targeting or profiling of individuals. Thus, these processes must incorporate human operators aside from AI-based prediction for double assurance. To address these issues, national security agencies should establish clear guidelines regarding the scope and limitations of surveillance, prioritize transparency wherever possible, and invest in unbiased AI models. This procedure will help ensure that surveillance efforts are fair and respect individuals' privacy rights.

6 Current Regulations and Frameworks

As AI-augmented OFS approaches evolve, the regulatory frameworks and ethical guidelines are crucial in ensuring their secure, ethical, and efficacious deployment. This section examines pivotal cybersecurity protocols and ethical frameworks relevant to AI-augmented OFS promoting data privacy, security, and integrity.

General Data Protection Regulation (GDPR): The GDPR, legislated by the European Union, is one among the most comprehensive data privacy regulations globally [16]. It mandates that organizations provide data protection, privacy, and transparency in handling personal and confidential data, including strict management of data access, storage, and processing. For AI-augmented OFS systems that assemble and process personal information, GDPR compliance is necessary to protect individuals' privacy and uphold data rights. This ensures that data collected by OFS sensors is securely stored, only accessible by authorized parties, and used transparently with consent.

Health Insurance Portability and Accountability Act (HIPAA): HIPAA is a U.S.-based ordinance that defends patient health information [17]. AI-driven OFS used for patient monitoring must adhere to HIPAA standards to ensure patient data remains private, secure, and conducted appropriately. This includes enforcing encryption, secure data transmission, and strict access control measures to protect sensitive medical data from unauthorized access or tampering.

NIST Cybersecurity Framework: Developed by the National Institute of Standards and Technology (NIST), this cybersecurity framework provides guidelines for managing cybersecurity risks [18]. It is widely embraced across sectors such as energy and transportation that use AI-augmented OFS in critical infrastructure monitoring. It emphasizes risk assessment, incident response, and continuous monitoring, enabling organizations to identify vulnerabilities and improve system resilience against cyberthreats proactively.

ISO/IEC 27001 and ISO/IEC 27017: These international standards concentrate on information security management and cloud security, respectively. It outlines best practices for protecting data integrity, confidentiality, and availability [19]. These standards apply to AI-augmented OFS systems that utilize cloud-based data processing. Compliance with ISO/IEC 27001 and 27017 can help organizations ensure their data handling, storage, and transmission practices meet rigorous security standards, reducing the risk of data breaches or unauthorized access.

IEEE's Ethically Aligned Design (EAD): The IEEE's EAD framework guides the ethical deliberations of AI and autonomous systems [20]. It accentuates transparency, accountability, and fairness. This framework facilitates and promotes the development of AI-augmented OFS systems, prioritizing privacy and minimizing bias, thus promoting fairness and inclusivity. In practical terms, EAD can guide organizations in implementing transparent AI decision-making processes in OFS applications, ensuring end users interpret, explain, and understand system outputs.

European Union (EU) AI Act: EU AI Act categorizes AI applications based on risk levels, with certain uses subject to higher scrutiny and compliance standards [21]. For AI-driven OFS applications in healthcare, critical infrastructure, and surveillance, compliance with this regulation will be necessary to address high-risk areas and ensure systems are designed and deployed responsibly. This act mandates high standards for data governance, risk management, and algorithm transparency, all essential for AI-augmented OFS systems handling sensitive data and impacting public welfare.

UNESCO's Recommendations on AI Ethics: UNESCO's guidelines on AI ethics concentrate on human rights, sustainability, and data privacy [22]. These suggestions stress the need for AI systems that admire individual rights, promote social welfare, and avoid harmful outcomes. For AI-powered OFS systems, adopting UNESCO's guidance encourages the responsible handling of environmental and public data, particularly in areas like environmental monitoring and public safety, where ethical considerations are vital.

7 Future Directions

OFS technology assisted by AI methods has huge potential across various industries. However, distinctive cybersecurity and ethical issues authenticate continuous investigation to extract its actual prospect. In the upcoming research, robust cybersecurity frameworks should be prioritized to counter risks like data interception, interpretation, unauthorized access, and AI-enabled OFS systems. It includes the development of resilient systems generated through the concurrence of encryption, anomaly detection, and secure AI training. Thus, generated systems will emerge as improved data protection and integrity in sectors involving healthcare and infrastructure.

Moreover, the ethical considerations in using AI methods in the OFS remain vital to extracting the utmost benefits. Therein, future research directions must explore and analyze the privacy-preserving AI training methods for privacy-critical applications. This ensures that the data is getting processed locally at the edge devices and maintains individual privacy and consent. To improve trust and system reliability, the enforcement of transparency and accountability mechanisms within AI methods are essential. Extending regulations, ethical frameworks, and guidelines tailored specifically for AI-assisted OFS can designate unambiguous standards. It paves the way for responsible and secure applications. Transistors [23, 24] accounting for high-speed and real-time signal processing (usually deployed in analog signal processing) are vital to secure the data and achieve efficient data processing. Thus, transistors play a crucial role in addressing ethical and security concerns to enhance the performance and reliability of analog circuits within these advanced systems. This critical function ensures that sensitive information transmitted through optical networks is safeguarded against potential threats, providing peace of mind in an increasingly interconnected digital landscape.

8 Conclusion

Integrating AI and OFS characterizes a substantial improvement across diverse industries, delivering enhanced accuracy, real-time data monitoring, and prediction abilities. Nevertheless, the advantages of AI-assisted OFS systems arrive with distinctive cybersecurity and ethical concerns. Assuring data safety and guarding against unauthorized access, interception, and AI-related threats is vital to successfully deploying these systems. Besides, ethical deliberations, fetching privacy protection, bias in AI methods, and the need for transparent, accountable decision-making are prominent in accumulating public trust and guaranteeing trustworthy and reliable use. This study highlights the significance of a double emphasis on technical advancements and the consequence of severe cybersecurity and ethical standards for AI-enabled OFS. With the technological evolution, continuing study and regulatory authorization will be necessary to handle these crises. To this end, the presented system will stimulate OFS to operate as a secure and ethical tool.

References

1. L. Alwis, T. Sun, K. Grattan, Developments in optical fibre sensors for industrial applications. Opt. Laser Technol. **78**, 62–66 (2016)
2. C. Pendão, I. Silva, Optical fiber sensors and sensing networks: overview of the main principles and applications. Sensors **22**(19), 7554 (2022)
3. B. Culshaw, Optical fibre sensors: A current perspective. Open Opt. J. **7**(1) (2013)
4. M. Abduljawwad, M. Khaleel, T.S. Ogedengbe, S. Abraheem, Sensors for daily utilization. Int. J. Electr. Eng. Sustain. 106–119 (2023)
5. L. Fanucci, S. Saponara, T. Bacchillone, M. Donati, P. Barba, I. SanchezTato, C. Carmona, Sensing devices and sensor signal processing for remote monitoring of vital signs in chf patients. IEEE Trans. Instrum. Meas. **62**(3), 553–569 (2012)
6. D. Dias, J. Paulo Silva Cunha, Wearable health devices—vital sign monitoring, systems and technologies. Sensors **18**(8), 2414 (2018)
7. M.A. Al Mamun, M.R. Yuce, Sensors and systems for wearable environmental monitoring toward IoT-enabled applications: A review. IEEE Sens. J. **19**(18), 7771–7788 (2019)
8. S.L. Ullo, G.R. Sinha, Advances in smart environment monitoring systems using IoT and sensors. Sensors **20**(11), 3113 (2020)
9. D.T. Delaney, G.M. O'Hare, A.G. Ruzzelli, Evaluation of energy efficiency in lighting systems using sensor networks, in *Proceedings of the First ACM Workshop on Embedded Sensing Systems for Energy-Efficiency in Buildings*, pp. 61–66 (2009)
10. M.N. Uddin, H. Nyeem, Engineering a multi-sensor surveillance system with secure alerting for next-generation threat detection and response. Results Eng. **22**, 101984 (2024)
11. C.-Y. Chong, S.P. Kumar, Sensor networks: evolution, opportunities, and challenges. Proc. IEEE **91**(8), 1247–1256 (2003)
12. M. Amanzadeh, S.M. Aminossadati, M.S. Kizil, A.D. Rakić, Recent developments in fibre optic shape sensing. Measurement **128**, 119–137 (2018)
13. M.H. Jarrahi, Artificial intelligence and the future of work: Human-AI symbiosis in organizational decision making. Bus. Horiz. **61**(4), 577–586 (2018)
14. U. Samal, A. Kumar, Enhancing software reliability forecasting through a hybrid Arima-ANN model. Arab. J. Sci. Eng. **49**(5), 7571–7584 (2024)

15. U. Samal, A. Kumar, A neural network approach for software reliability prediction. Int. J. Reliab. Qual. Saf. Eng. 2450009 (2024)
16. C.J. Hoofnagle, B. Van Der Sloot, F.Z. Borgesius, The european union general data protection regulation: what it is and what it means. Inf. Commun. Technol. Law **28**(1), 65–98 (2019)
17. A. Act, Health insurance portability and accountability act of 1996. Public Law **104**, 191 (1996)
18. B. Krumay, E.W. Bernroider, R. Walser, Evaluation of cybersecurity management controls and metrics of critical infrastructures: A literature review considering the NIST cybersecurity framework, in *Secure IT Systems: 23rd Nordic Conference, NordSec 2018, Oslo, Norway, November 28–30, 2018, Proceedings 23*, (Springer, 2018), pp. 369–384
19. C. Mitchell, Privacy, compliance and the cloud. Guid. Secur. Assur. Cloud Comput. 3–14 (2015)
20. K. Shahriari, M. Shahriari, IEEE standard review—ethically aligned design: A vision for prioritizing human wellbeing with artificial intelligence and autonomous systems, in *2017 IEEE Canada International Humanitarian Technology Conference (IHTC)*, (IEEE, 2017), pp. 197–201
21. L. Edwards, The EU AI act: A summary of its significance and scope. Artif. Intell. (the EU AI Act) **1** (2021)
22. D.E. van Norren, The ethics of artificial intelligence, UNESCO and the african ubuntu perspective. J. Inf. Commun. Ethics Soc. **21**(1), 112–128 (2023)
23. A. Das, S. Rewari, B.K. Kanaujia, S.S. Deswal, R.S. Gupta, Physics based numerical model of a nanoscale dielectric modulated step graded germanium source biotube FET sensor: Modelling and simulation. Phys. Scr. **98**(11), 115013 (2023). https://doi.org/10.1088/1402-4896/acf4c9
24. A. Das, S. Rewari, B.K. Kanaujia, S.S. Deswal, R.S. Gupta, Analytical modeling and doping optimization for enhanced analog performance in a Ge/Si interfaced nanowire MOSFET. Phys. Scr. **98**(7), 74005 (2023). https://doi.org/10.1088/1402-4896/acde16

Hybrid Systems Combining Traditional and AI-Enhanced Optical Fiber Sensors

Arvind Sharma, Sugandha Gupta, and Kavita Rani Segwal

Abstract Integrating AI with optical fiber sensors allows a hybrid system that enhances strength for detecting various physical parameters such as intensity, data transmission, wavelength, and phase. These hybrid systems have been widely used in environmental sensing, telecommunication, and health monitoring. By analyzing the previous and current data trends, AI models enable predictive future maintenance, which reduces interruption and maintenance costs. These systems are better known for their high sensitivity, improved work efficiency in harsh environments, wide dynamic range, and are immune to electromagnetic (EM) interference and electrical noise. This integration minimizes environmental impact, reduces downtime, increases lifetime benefits, improves resource management, and optimizes production. This chapter elucidates the hybrid optical fiber sensors, hybrid system strengths, limitations, applications, and future trends.

Keywords Optical Fibre Sensors (OFS) · Artificial Intelligence (AI) · Smart city

1 Introduction

Industry 4.0 has significantly accelerated the development of hybrid-based sensors. It plays a pivotal role in the diverse application sectors of industrial processes [1], biomedical engineering [2, 3], robotics [4], civil engineering [5], and digitalization and intelligence in the health monitoring sector [6]. These sensors are made from advanced materials and have sophisticated designs and sensitive units to transform stimuli from objects into electrical or optical signals. These signals are further processed through filtering, impedance matching, amplification, and interoperability. Moreover, Industry 4.0, the best feature for integrating AI with an optical fiber sensor, has become a key technology in health monitoring systems [7]. Combining AI with an

A. Sharma (✉) · S. Gupta · K. R. Segwal
Department of Applied Science (Physics), Bhagwan Parshuram Institute of Technology, Delhi, India
e-mail: arvindsharma1900@gmail.com

V. Bhardwaj et al. (eds.), *Optical Fiber Sensors and AI*, Progress in Optical Science and Photonics 34, https://doi.org/10.1007/978-981-96-6158-9_11

optical fiber sensor has numerous benefits, such as accuracy, data transmission, durability and reliability, and real-time monitoring. Hence, this hybrid system in industry automation allows smooth operation by detecting early defects from production lines and reducing repair costs and manual interference, enhancing production efficiency and product quality [8]. Sensors with high accuracy and multifunctional adaptability are required in the field of robotics, such as unmanned aerial vehicles [9] and deep-sea robots [10], which can adjust in dynamic environments and perform required tasks [11]. Similarly, sensors in the health monitoring sector detect subtle physiological or structural changes with high precision and reliability under complex conditions [12]. For example, railway responses in permafrost regions were monitored [13], and physiological data on human skin during motion was collected [14]. Hence, introducing advanced AI algorithms, such as machine learning and deep learning technologies, plays a crucial role in overcoming these traditional trends [15]. They enhance sensor accuracy [16] and sensitivity [17] under definite conditions and boost adaptability [18] to environmental shifts. These technologies not only monitor but also enable efficient identification and predictive capabilities in Industry 4.0 [19], disease diagnosis [20], the environmental awareness and adaptability of robots [21], and structural damage prevention [22]. Current research focuses on integrating artificial intelligence [23] with sensor technology, from enhancement algorithms [24] to AI-driven device designs [25], with broad applications in biomedical [26] and engineering fields [27]. In this book chapter, we delve into the vital functions of artificial intelligence algorithms in optical sensing, highlighting the latest progress in innovative applications. This chapter discusses the concept of hybrid optical fiber sensors and hybrid system strengths, limitations, applications, and future trends.

2 Fundamentals of Optical Fiber Sensors

Optical fiber sensors work on light propagation, an excellent invention for detecting various physical parameters such as intensity, data transmission, wavelength, and phase. These small and economical sensors work better in harsh environments, have a wide dynamic range, and are immune to electromagnetic interference and electrical noise. Because of these features, they replaced old conventional sensors. A typical fiber sensor block diagram is shown below.

The block diagram shown in Fig. 1 has the following key parameters: the optical source, sensing element, optical detector, and processing unit. The typical optical sources that produce the light are LED and laser, respectively, which are transmitted through an optical fiber to the sensing element. This fiber acts as a pathway to guide the light wherever it needs to go. The sensing element or transducer is the platform used to measure the various physical parameters of data. The optical detector detects these signals and then converts them into electrical signals. These electrical signals are then fed into the processing unit. Lastly, the processing unit uses an oscilloscope or analyzer to process these electrical signals into valuable data such as the detected

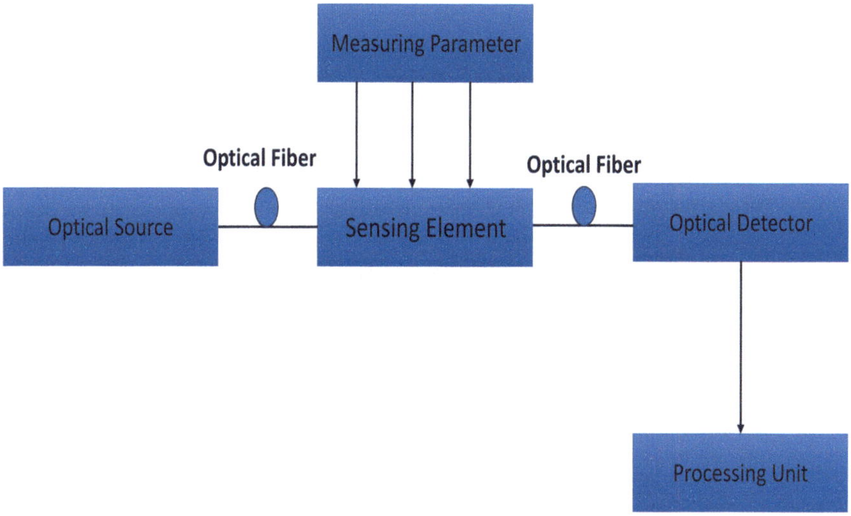

Fig. 1 Block diagram of optical fiber sensor

intensity, temperature, and physical parameter change. Hence, each parameter plays a crucial role in the better performance and accuracy of optical fiber sensors [28, 29].

2.1 Optical Fiber Sensors (OFS) Categories

OFS are divided into three types:

1. **Sensing Location** is further divided into intrinsic and extrinsic sensors [30].
   (a) **Intrinsic**: In this type of sensor, sensing directly occurs in the fiber. This is because the environment affects the physical parameters such as strain, pressure, and temperature and changes the optical properties such as phase or intensity. They are explicitly designed to measure over a long distance, making them a suitable candidate for pipeline surveillance or structural monitoring.
   (b) **Extrinsic**: This type of sensor senses outside the fiber, with the fiber acting to carry light to and from the sensing area. It acts as an information carrier that transmits light to a remote receiver using a mirror. The benefit of this sensor is that it is accessible to remote locations. Extrinsic sensors measure vibration, rotation, torque, and twisting.
2. **Operating Principle**: On this basis, sensors are classified into intensity, wavelength, phase, and polarization-based sensors [28, 31].

(a) **Intensity-Based Sensing**: It detects surrounding changes such as strain, temperature, and pressure by measuring the change in light intensity transmitted through an optical fiber. This type of sensor is more straightforward and, therefore, has a lower cost. This type of sensor is not used because anything that causes a change in optical power will generate a reading. Thus, it is difficult to eliminate noise and bogus readings from these sensors.
(b) **Phase-Based Sensing**: It detects changes in variables like displacement, vibration, or acoustic changes using the phase shift of light waves in the fiber. The Fabry–Perot interferometers and Mach–Zehnder Michelson interferometer are the most popular. This type of sensor offers high precision and is widely used in seismic monitoring.
(c) **Wavelength-Based Sensing**: It uses interferometric sensors or Fiber Bragg gratings to detect temperature, strain, or chemical variations. This type of sensor relies on spectral shift. These sensors are embedded in large structures like dams or bridges. An example of wavelength-based sensing is Fiber Bragg Gratings (FBGs), which reflect specific wavelengths that shift based on external factors.
    1. **Polarization-Based Sensing**: The light polarization concept measures current and strain to perform biomedical sensing. This type of sensor is used to sense electric current and follow the Faraday effect, which produces a rotation in the polarization plane of light in the presence of a magnetic field.

3. **Application-Based OFS**: It depends on the physical measurement, mainly strain, pressure, and temperature. For example, vital signs on active farm animals are monitored, including body temperature, fluid pressure, and blood flow, respectively. The chemical sensors detect pH, gas composition, or glucose levels. Biomedical sensors are used in medical diagnostics, including blood glucose monitoring, oxygen level detection, and nerve activity mapping. Some examples of optical fiber sensors are described below [29, 32, 33].

    1. **Intensity-Based Sensors**: These sensor designs are simple but limited to environmental noise. This sensor detects any physical disturbance caused by the intensity change.
    2. **Interferometric Sensors**: These precisely measure path length variations and utilize the interference pattern generated from coherent light sources.
    3. **Fiber Bragg Gratings**: The sensor geometry is periodic and reflects only a specific wavelength. The shift in wavelength allows precise sensing due to changes in strain or temperature.

2.2 Limitations

Optical fiber sensors have several advantages, but they still have some flaws [29].

1. **Signal Degradation**: Over a significant distance, signals get weakened, and data clarity is lost.
2. **Susceptibility to External Factors**: It is sensitive to any change in external factors, such as temperature and vibration, that interfere with measurement accuracy, such as noise.
3. **Difficulty in Interpreting Data**. The information collected from optical fiber sometimes becomes complex and challenging to interpret, requiring specialized knowledge.

These limitations make it harder to obtain actionable insights from the collected data.

3 Role of Artificial Intelligence in Sensor Systems

Integrating Artificial intelligence (AI) with optical fiber sensors is crucial in Industry 4.0. [34] With the advent of cutting-edge technology, it is necessary to adopt innovative systems. Integrating AI with traditional sensors is vital for optimization, predictive maintenance, and better decision-making. By combining optical fiber sensors, we can establish a network of sensors along a single fiber that is expected to conduct routine inspections and identify structural deterioration very early [35]. The benefits of OFS are as follows:

(a) **Accuracy**: It precisely measures critical infrastructure.
(b) **Data Transmission**: Data is transmitted for distances without significant loss.
(c) **Durability and Reliability**: It can resist harsh environmental conditions, confirming long-term reliability.
(d) **Live Monitoring**: It allows for informed decisions.

In smart cities, sensors generate large amounts of data, which is difficult to handle by a conventional approach. Hence, AI integration become pivotal as their primary function is data analysis. AI uses algorithms such as machine and deep learning to identify anomalies and patterns and any infrequent changes that are not obvious initially. For example, in manufacturing, AI is used to examine parameters such as temperature, vibration, and pressure, allowing timely information on which parts are not working correctly, fixing on time, and reducing escalating costs. It increases the predictive capability of the optical fiber system. Using the predictive approach, AI used the previous data reading with the current reading to show which equipment part would not work correctly. These strategies allow companies to go from reactive to proactive maintenance, reducing maintenance timing and cost. For example, in the aerospace industry, AI-based sensors can track engine performance and predict which equipment will likely fail, allowing scheduled maintenance and reduced operational distributions. Also, AI-integrated sensors in smart factories allow optimized production schedules to manage the supply chain and stock operation. This will increase efficiency and flexibility and help them quickly respond to demand and production conditions. Integrating AI sensors improves the accuracy and reliability of measurement. It filters out unnecessary noise, ensuring proper decision-making, particularly in healthcare, where reading makes a big difference in patient health, creating an innovative environment, adjusting real-time security systems, and optimizing energy consumption [36–39]. Integrating AI with an optical fiber sensor helps in several ways.

(a) **Noise Reduction**: AI algorithms help to filter out background noise and signal data for better measurement accuracy. However, crucial environmental factors, such as vibrations, fluctuation in temperature, and EM radiation interference, can distort the signal from the sensor reading.
(b) **Pattern Recognition**: AI machine learning helps to track even the extensive, noisy data from the sensor using a pattern approach. This allows early fault detection for environmental anomalies and structural damage.
(c) **Predictive Analysis**: By analyzing the previous and current reading trends, AI models enable predictive future maintenance, which reduces interruption and maintenance costs.

Furthermore, it allows us to overcome many challenges through noise reduction and better decision-making processes. It can handle complex tasks to enhance accuracy by spotting even small changes.

4 Hybrid Systems: Integration of AI with Optical Fiber Sensors

Integrating AI with an optical fiber sensor [40] allows a hybrid system that enhances industry and health sector strength. It improves the accuracy of performance functionality among various applications. Optical fiber sensors are known for their high sensitivity, work better in harsh environments, have wide dynamic range, and are immune to EM interference and electrical noise. It allows real-time data analysis for faster and smoother operation of complex datasets that lead to unwanted interference and actionable insight for predictive capability and better decision-making [39]. Various approaches can help achieve these applications.

(a) **Data Acquisition and Preprocessing**: The AI algorithm uses raw data from an optical fiber sensor to reduce the signal-to-noise ratio and applies a filter technique to obtain relevant data.
(b) **Feature Extraction**: It is used to extract the feature from sensor data, which is essential for determining the physical parameter for monitoring. To identify such features, Machine learning algorithms allow better interpretation of sensor signals.
(c) **Model Deployment**: Machine learning models are used for large datasets to identify the relationship between sensor data and physical phenomena and enable accurate predictions and diagnosis.

(d) **Real-Time Monitoring**: Integrating AI sensors allows real-time monitoring by analyzing sensor data. It will enable the generation of alerts for timely action when critical thresholds are breached.

For example, monitoring the condition of concrete structures, specially reinforced with textile-reinforced concrete optical fiber sensor integration, becomes essential and has benefits such as resistance to high temperatures, multiplexing capability, and immunity to EM interference. Hence, it is ideal for detecting issues of concrete buildings. By integrating optical fiber sensors, we may establish a network of sensors along a single fiber, conduct routine inspections, and identify structural deterioration early [35]. Hence, OFS becomes an essential parameter in innovative infrastructure development, providing live monitoring for health infrastructure. Their integration with other smart systems helps us monitor structural health, detect anomalies, and safeguard infrastructure integrity. Optical fiber sensors actively help by providing on-time information about physical parameters, such as strain, temperature, and vibrations, enhancing the smart infrastructure's ability to prevent failure. Hence, its accurate and reliable measurement makes the smart infrastructure requisite for confirming longevity and resilience and can withstand harsh environmental conditions. The continuous evolution of optical fiber sensor technology shows potential for monitoring and managing smart infrastructures.

5 Advantages of Hybrid Systems

Integrating AI with optical fiber sensors forms a hybrid system for various technological sensor applications [36, 39, 41, 42]. These hybrid systems have many key advantages, as discussed below.

(a) **Sensor Data Analysis**: Hybrid systems use advanced AI algorithms to process and analyze massive sensor data. This allows us to identify the hidden trend and pattern that conventional analysis might miss, leading to more precise decision-making.
(b) **Predictive Maintenance**: AI helps to analyze old and real-time sensor data trends to predict potential failures in equipment before they occur. This proactive maintenance strategy prevents unexpected downtime, reduces repair costs, and increases the equipment's lifetime.
(c) **Accuracy and Reliability**: The AI algorithm reduces the signal-to-noise ratio and applies a filter technique to obtain relevant data. It is a critical application in health sectors and can make a big difference between life and death.
(d) **Real-Time Monitoring**: Hybrid systems allow real-time monitoring of various physical parameters for proactive response to any immediate change in condition. For example, in an industrial factory, AI collects real-time sensor data, optimizes production efficiency, and reduces material waste.

(e) **Smart Decision-Making**: Integrating AI enables automated decision-making based on sensor data and allows resource management, optimized production schedule, and productivity. This helps to manage complex and harsh environmental conditions.
(f) **Adaptability and Learning**: A hybrid system can learn from old data and adapt to new conditions, making it more flexible and able to perform in dynamic environments. For example, in remote areas, air pressure and temperature parameters change dramatically; in that case, the AI-enabled system proactively helps identify the data trends and provides actionable insights to overcome natural calamities.
(g) **Cost Efficiency**: A hybrid system reduces the manual operating and maintenance cost. This feature allows the industry to avoid production downtime and repair costs.
(h) **Scalability**: A hybrid system is easily scaled for a wide range of monitoring and requires additional sensors to meet growing data needs. This has become essential for industries like smart cities and telecommunications.
(i) **Enhanced Safety**: Hybrid systems can efficiently operate in harsh environments, which reduces the risk to human operators. For example, optical fiber sensors can be placed in hazardous environments while AI analyzes these data to ensure safety.
(j) **Cross-Industry Applications**: The flexibility of hybrid systems makes them good candidates in various sectors, such as environmental monitoring, healthcare, manufacturing, and transportation. This broad application makes it a valuable tool to address diverse challenges.

6 Case Studies

For a combined hybrid system the following examples are discussed below:

(a) **Structural Health Monitoring**: Monitoring the structural integrity of bridge hybrid systems plays a crucial role in determining fault and fracture. These sensors are embedded in concrete to detect strain and temperature changes. Sensor data is collected and transmitted to AI advanced algorithm models such as machine learning to analyze such data for the early detection of potential failure. This will help us monitor structural health, detect anomalies, and safeguard infrastructure integrity, enabling timely alerts when critical thresholds are breached [35].
(b) **Biomedical Monitoring**: In the biomedical field, the amalgamation of optical fiber sensors with AI algorithms monitors glucose levels in diabetic patients. Also, changes in the illumination absorption coefficient in the blood can be detected with these smart fiber detectors, which help doctors analyze and correlate glucose concentration. Machine learning techniques provide real-time glucose monitoring, which allows for proactively determining the requirement

of buffer blood bottles in an emergency and accurate and non-invasive glucose level tracking [26].

(c) **Environmental Monitoring**: Hybrid sensors powered with AI are used to detect water pollution levels. A data set is prepared between light transmission coefficient and different contaminants in water. Later, an ML algorithm is deployed to identify types and percentage levels of pollutants based on recorded sensor data. Due to this, real-time monitoring of environmental hazards can be predicted and controlled [21].

(d) **Smart Agriculture**: Agriculture is predominantly responsible for accelerating a country's economic growth. Therefore, AI optical sensors provide valuable information regarding soil conditions, pest development, and irrigation needs. This hybrid technique has improved the quality and quantity of production with optimized water usage and better waste management [43].

7 Challenges and Limitations

Integrating AI with an optical fiber sensor forms a hybrid system with many key advantages. However, some issues still open broad research areas for further study. The primary concerns are data quality and quantity. It is difficult to get good-quality data, as limited lab data can cause overfitting. The model also ensures it works well on unseen data, and the adaptability of the sensor with time is also a crucial factor for reliability. The other major issue is the power consumption of the device. For example, executing a deep learning model on the multisensory chip while maintaining real-time response and accuracy is challenging. Optimized sensor performance for diverse application model interpretability is critical to understanding the AI decision-making process [34, 39–41]. The other limitations are:

1. **Complexity of Integration**: Integrating AI with optical fiber sensors with advanced algorithms requires complex system design, which includes interoperability between the AI platform and device hardware. Addressing such bottleneck issues entails multidisciplinary expertise.
2. **Noise**: Optical fiber sensors mainly work in noisy environments such as high-temperature and high-vibration situations. Hence, the poor quality of the sensing data reduces the model's performance compared to noise filtering and preprocessing methods, which need further enhancement.
3. **Ethical Challenges**: Hybrid systems lead to many ethical concerns, including smart decision-making processes with transparency by mitigating biases in hybrid systems. Strict sensor regulator standards delay the deployment timeline and complicate the system design.

8 Future Prospect

The following key points focus on the future trend of hybrid systems [34, 39].

1. **Advanced Algorithms for Data Processing**: Integrating AI models with optical fiber sensor data has become a focal point. Techniques like transfer learning and federated learning improved AI performance even with limited or scattered datasets, keeping data privacy.
2. **Real-Time Edge Computing Analysis**: Integrating edge computing into hybrid systems can cut delay and enhance real-time processing. This is beneficial for remote monitoring and smart city applications.
3. **Self-Powered**: Research on energy-harvesting technologies, such as piezoelectric or triboelectric, is growing in demand. Thus, it works without an external battery, enabling continuous and autonomous operation in remote areas.
4. **Hybrid AI Models**: Integrating physics-based models with AI provides new opportunities regarding future aspects. These combine knowledge of optical systems with AI, improving the reliability and accuracy of hybrid systems.
5. **Fusion of Multi-Modal Sensor**: Various sensor integrations, such as optical, environmental, and acoustic, provide a complete view of monitored environments.
6. **IoT and Cloud Integration**: The remote area monitoring and control hybrid system utilizes IoT and the cloud platform. Hence, cloud-based AI helps the sensor learn over time.
7. **Advanced Materials**: Using novel materials such as polymers or nanostructures-based optical fiber sensors will improve durability and sensitivity. Hence, integrating with AI enhances performance and achieves sustainability.
8. **Reliability and Transparency**: The hybrid system becomes questionable for transparent and reasonable decision-making processes, raising concerns about accountability and ethical considerations. Thus, strict regulatory compliance is needed for better user trust.
9. **Emerging Fields Applications**: Hybrid systems have become essential for disaster monitoring, autonomous vehicles, and biomedical sensing, where smart data processing and high precision are crucial.

9 Conclusion

Integrating AI with optical fiber sensors is a transformative step in Industry 4.0. It has many benefits, such as reducing the number of sensors, which improves precision, reliability, and scalability. This integration minimizes environmental impact, improves resource management and predictive maintenance, and improves decision-making by using AI to process complex data efficiently. This allows real-time monitoring for early warning and fault detection, predictive analysis, and noise reduction for various applications to address modern challenges such as manufacturing, healthcare, smart agriculture, and environmental monitoring. Despite specific challenges like noise interference and integration complexity, algorithms, materials, and energy efficiency advancements have been of a great deal of interest in developing a sustainable future. This hybrid system concept led to sustainable development, ensuring resilience and efficiency in harsh environments and remote areas.

References

1. L. Spinelle, M. Gerboles, G. Kok, S. Persijn, T. Sauerwald, Review of portable and low-cost sensors for the ambient air monitoring of benzene and other volatile organic compounds. Sensors **17**, 1520 (2017)
2. D. Luo, H. Sun, Q. Li, X. Niu, Y. He, H. Liu, Flexible sweat sensors: From films to textiles. ACS Sens. **8**, 465–481 (2023)
3. J. Gu, Y. Shen, S. Tian, Z. Xue, X. Meng, Recent advances in nanowire-based wearable physical sensors. Biosensors **13**, 1025 (2023)
4. J. Chen, Y. Zhu, X. Chang, D. Pan, G. Song, Z. Guo, N. Naik, Recent progress in essential functions of soft electronic skin. Adv. Funct. Mater. **31**, 2104686 (2021)
5. Y. Bao, Z. Chen, S. Wei, Y. Xu, Z. Tang, H. Li, The state of the art of data science and engineering in structural health monitoring. Engineering **5**, 234–242 (2019)
6. Z. Zhang, L. Wang, C. Lee, Recent advances in artificial intelligence sensors. Adv. Sens. Res. 2200072 (2023)
7. M. Huang, M. Zhu, X. Feng, Z. Zhang, T. Tang, X. Guo, T. Chen, H. Liu, L. Sun, C. Lee, Intelligent cubic-designed piezoelectric node (iCUPE) with simultaneous sensing and energy harvesting ability toward self-sustained artificial intelligence of things (AIoT). ACS Nano **17**, 6435–6451 (2023)
8. T. Sarang, Z. Zheng, R. Henrik, M. Jose, M. Tuomo, N. Sergey, H. Toni, N.M. Hiski, H.B. Zahidul, V.L. Ville et al., Sensors and AI techniques for situational awareness in autonomous ships: A review. IEEE Trans. Intell. Transp. Syst. **23**, 64–83 (2022)
9. H. Shakhatreh, A.H. Sawalmeh, A. Al-Fuqaha, Z. Dou, E. Almaita, I. Khalil, N.S. Othman, A. Khreishah, M. Guizani, Unmanned aerial vehicles (UAVs): A survey on civil applications and key research challenges. IEEE Access **7**, 48572–48634 (2019)
10. L. Guorui, C. Xiangping, Z. Fanghao, L. Yiming, X. Youhua, C. Xunuo, Z. Zhen, Z. Mingqi, W. Baosheng, Y. Shunyu et al., Self-powered soft robot in the Mariana trench. Nature **591**, 66–71 (2021)
11. W. Yuan, S. Dong, E.H. Adelson, GelSight: High-resolution robot tactile sensors for estimating geometry and force. Sensors **17**, 2762 (2017)
12. K. Yasser, E. Aminy, M. Claire, P. Adrien, C. Ana, Monitoring of vital signs with flexible and wearable medical devices. Adv. Mater. **28**, 4373–4395 (2016)
13. W. Ma, G. Cheng, Q. Wu, Construction on permafrost foundations: Lessons learned from the Qinghai-Tibet railroad. Cold Reg. Sci. Technol. **59**, 3–11 (2009)
14. H. Song, G. Luo, Z. Ji, R. Bo, Z. Xue, D. Yan, F. Zhang, K. Bai, J. Liu, X. Cheng et al., Highly-integrated, miniaturized, stretchable electronic systems based on stacked multilayer network materials. Sci. Adv. **8**, eabm3785 (2022)
15. F. Zhou, Y. Chai, Near-sensor and in-sensor computing. Nat. Electron. **3**, 664–671 (2020)
16. F. Wen, Z. Sun, T. He, Q. Shi, M. Zhu, Z. Zhang, L. Li, T. Zhang, C. Lee, Machine learning glove using self-powered conductive superhydrophobic triboelectric textile for gesture recognition in VR/AR applications. Adv. Sci. **7**, 2000261 (2020)
17. H.J. Young, K.H. Seong, S.W. Hee, H.H. Jae, X. Trung, P. Hyunsin, K. Junyeong, K. Sunghun, D.Y. Chang, J.L. Keon, Flexible piezoelectric acoustic sensors and machine learning for speech processing. Adv. Mater. **35**, 1904020 (2020)

18. S. Muhammad, B. Stephan, S. Hans, J.M.H. Paul, D.I. Ozlem, Towards detection of bad habits by fusing smartphone and smartwatch sensors, in *Proceedings of the 2015 IEEE International Conference on Pervasive Computing and Communication Workshops (PerComWorkshops)* (St. Louis, MO, USA, 23–27 March 2015), pp. 591–596
19. Z. Chen, W. Li, Multisensor feature fusion for bearing fault diagnosis using sparse autoencoder and deep belief network. IEEE Trans. Instrum. Meas. **7**, 1693–1702 (2017)
20. S.H. Kwon, L. Dong, Flexible sensors and machine learning for heart monitoring. Nano Energy **102**, 107632 (2022)
21. S. Chun, J.S. Kim, Y. Yoo, Y. Choi, S.J. Jung, D. Jang, G. Lee, K.I. Song, K.S. Nam, I. Youn et al., An artificial neural tactile sensing system. Nat. Electron. **4**, 429–438 (2021)
22. M. Faiz, S. Ramandeep, A. Mohd, A.S. Anwar, B. Chandani, F. Nausheen, Machine learning techniques in wireless sensor networks: Algorithms, strategies, and applications. Int. J. Intell. Syst. Appl. Eng. **11**, 685–694 (2023)
23. Y. Wang, M.L. Adam, Y. Zhao, W. Zheng, L. Gao, Z. Yin, H. Zhao, Machine learning-enhanced flexible mechanical sensing. Nano-Micro Lett. **15**, 190–222 (2023)
24. A. Vergara, S. Vembu, T. Ayhan, M.A. Ryan, M.L. Homer, R. Huerta, Chemical gas sensor drift compensation using classifier ensembles. Sens. Actuators B Chem. **166–167**, 320–329 (2012)
25. X. Cheng, Z. Fan, S. Yao, T. Jin, Z. Lv, Y. Lan, R. Bo, Y. Chen, F. Zhang, Z. Shen et al., Programming 3D curved mesosurfaces using microlattice designs. Science **379**, 1225–1232 (2023)
26. A. Moin, A. Zhou, A. Rahimi, A. Menon, S. Benatti, G. Alexandrov, S. Tamakloe, J. Ting, N. Yamamoto, Y. Khan et al., A wearable biosensing system with in-sensor adaptive machine learning for hand gesture recognition. Nat. Electron. **4**, 54–63 (2020)
27. J. Yang, J. Chen, Y. Su, Q. Jing, Z. Li, F. Yi, X. Wen, Z. Wang, Z.L. Wang, Eardrum-inspired active sensors for self-powered cardiovascular system characterization and throat-attached anti-interference voice recognition. Adv. Mater. **27**, 1316–1326 (2015)
28. C. Pendão, I. Silva, Optical fiber sensors and sensing networks: Overview of the main principles and applications. Sensors **22**, 7554 (2022)
29. W. Lin, C. Zhang, L. Li, S. Liang, Review on development and applications of fiber-optic sensors. in *Proceedings of the 2012 Symposium on Photonics and Optoelectronics* (Shanghai, China, 21–23 May 2012), pp. 1–4
30. S.E.U. Lima, O. Frazão, F.M. Araujo, L.A. Ferreira, V. Miranda, J.L. Santos, Extrinsic and intrinsic fiber optic interferometric sensors for acoustic detection in high-voltage environments. Opt. Eng. **48**(2), 024401
31. D.A. Krohn, T. MacDougall, A. Mendez, *Fiber optic sensors: fundamentals and applications; Spie Press: Bellingham* (WA, USA, 2014)
32. Mittal, S.; Sharma, T.; Tiwari, M. Surface plasmon resonance based photonic crystal fiber biosensors: A review. Mater. Today Proc. 2021, 43, 3071–3074. [CrossRef]
33. M.D. Marazuela, M.C. Moreno-Bondi, Fiber-optic biosensors—An overview. Anal. Bioanal. Chem. **372**, 664–682 (2002)
34. L. Chen, C. Xia, Z. Zhao, H. Fu, Y. Chen, AI-driven sensing technology: Review. Sensors **24**(2024)
35. L.S.M. Alwis, K. Bremer, B. Roth, Fiber optic sensors embedded in textile-reinforced concrete for smart structural health monitoring: A review. Sensors **21**, 4948 (2021)
36. G. Valente, V. Muttillo, M. Muttillo, G. Barile, A. Leoni, W. Tiberti, L. Pomante, SPOF—slave powerlink on FPGA for smart sensors and actuators interfacing for industry 4.0 applications. Energies **12**(9), 1633 (2019)
37. M. Silva, E. Vieira, G. Signoretti, I. Silva, D. Silva, P. Ferrari, A customer feedback platform for vehicle manufacturing compliant with industry 4.0 vision. Sensors **18**(10), 3298 (2018)
38. K. Mahato, B. Purohit, A. Kumar, P. Chandra, Clinically comparable impedimetric immunosensor for serum alkaline phosphatase detection based on electrochemically engineered Au-nano-Dendroids and graphene oxide nanocomposite. Biosens. Bioelectron. **148**, 111815 (2020)

39. M. Javaid et al., Significance of sensors for industry 4.0: Roles, capabilities, and applications. Sens. Int. **2**, 100110 (2021)
40. Y. Zhang et al., Fully automated and AI-assisted optical fiber sensing system for multiplexed and continuous brain monitoring. ACS Sens. **9**, 6605–6620 (2024)
41. A. Schütze, N. Helwig, T. Schneider, Sensors 4.0–smart sensors and measurement technology enable Industry 4.0. J. Sens.S Sens. Syst. **7**(1), 359–371 (2018)
42. F. Ali, S. El-Sappagh, S.R. Islam, A. Ali, M. Attique, M. Imran, K.S. Kwak, Intelligent healthcare monitoring framework using wearable sensors and social networking data. Futur. Gener. Comput. Syst. **114**, 23–43 (2021)
43. S. Qazi, B.A. Khawaja, Q.U. Farooq, IoT-equipped and AI-enabled next generation smart agriculture: A critical review, current challenges, and future trends. IEEE Access **10**, 21219–21235 (2022)

Future Perspectives: Innovations and Challenges in AI-Powered Optical Fiber Sensing

Amit Agarwal and Monu Nath Baitha

Abstract Artificial intelligence (AI) techniques are currently gaining momentum in every corner of the study and analysis, including optical fiber sensing. These techniques are more optimal and robust for designing unprecedented levels of precision and functionality in optical fiber sensing system. This chapter provides a clear picture of different steps and operations in designing AI-powered optical sensor environment. The varying sensor data from different sensor type has different characteristics. This chapter clearly defines the characteristics of different types sensor datasets and suitable pre-processing steps for converting the dataset suitable for AI model learning. Moreover, important feature extraction methods are also discussed which are different for each optical sensor datasets and essential in determining patterns and relationships in sensor data. In subsequent part of the chapter AI model selection methodology, evaluation metrics selection for different analysis and data types with data augmentation methods and hyperparameter optimization methods are clearly devised for optimal decision-making in the field of optical fiber sensors. This multi-dimensional analysis can act as a guideline for future researchers and practitioners to navigate the complexities and harness the full potential of AI-powered optical fiber sensing.

Keywords Artificial intelligence · Optical fiber sensing · Optical fiber sensors data types · Feature extraction · Data augmentation

A. Agarwal
Department of Electrical and Electronics Engineering, ABES Engineering College, Ghaziabad, UP, India

School of Computer Science & Engineering, IILM University, G. Noida, India

M. N. Baitha (✉)
Department of Information Engineering, University of Brescia, Via Branze 38, 25123 Brescia, Italy
e-mail: monu.baitha@unibs.it

V. Bhardwaj et al. (eds.), *Optical Fiber Sensors and AI*, Progress in Optical Science and Photonics 34, https://doi.org/10.1007/978-981-96-6158-9_12

1 Introduction

Many developments have taken place in the field of optical fiber sensing as it possesses unique characteristics like high sensitivity, immunity to electromagnetic interferences, and its functioning capabilities under harsh environments [1]. These features have made optical fiber sensors a fundamental tool in many such applications as structural health monitoring, environmental sensing, and biomedical diagnostics. But as sensors become ever more ubiquitous and the data generated by them becomes larger and more complex, traditional data analysis methods are starting to seep into their limits [2]. As a consequence, there is an increasing demand to integrate artificial intelligence (AI) with optical fiber sensing in order to improve performance and enable new functionalities [3].

Industries from healthcare and finance to autonomous systems and natural language processing have been revolutionized by AI. This makes it a useful tool for improving sensor performance, as the platform can process large datasets and discover trends or make real-time decisions [4]. The main advantage of AI techniques lies in their capacity of handling massive datasets and handling real-time inputs which facilitates the decision-making capacity to the optical fiber sensing operators. Whereas the tradition optical fiber sensing methods lagged in efficiency and speed compared to the contemporary AI techniques. The availability of large variety of AI models also helps in selecting the model according to the requirement and characteristic of the dataset which could be different for different types of sensors [5, 6]. Different types of AI models already reported in the optical fiber sensing are linear regression, support vector machine, random forest, and neural network, etc.

Optical fiber sensors work in the real-time application environment. These operating conditions of the sensor could lead to noisy data generation, and often the resolution of the optical sensor datasets is very low which generates massive amount of the data. Utilization of traditional data analysis techniques such as Fourier series was utilized but it often struggles with real-time, large-scale data processing and adaptive learning methods [7, 8]. AI power data handling techniques in this pre-processing step automate the process which improves accuracy and reliability by learning from past data, adjusting to new conditions, and processing information in real-time without requiring manual input. These early detection of structural issues in buildings or bridges enables timely maintenance and minimizes the risk of catastrophic failures.

The application of AI in optical fiber sensing can revolutionize different aspects of working and automation [9]. The smart city concept lies in the automation which largely depends on different optical fiber sensors and its handling in real-time operation and control. Applications such as traffic pattern identification and control, optical fiber sensors offer real-time insight into infrastructure conditions, detect leaks in water and gas pipelines, and monitor and control air quality in smart cities can make the standard of living to more viable [10, 11].

Likewise, in healthcare, AI based optical fiber sensors are creating a new landscape for more sophisticated diagnostic tools and monitoring systems. These sensors with AI model could analyze data from these sensors to track vital signs, detect early disease indicators, and monitor the development of chronic illnesses. These innovative technologies are growing very fast in wearable electronic devices as it monitors the vital parameter of the humane being in real time and alarm the user in extreme health conditions thereby saving the humane lives [12, 13].

Even with these improvements, integrating AI and optical fiber sensing still faces some hurdles. One big problem is getting enough good data to train machine learning models. It's often hard to find this data for rare or extreme situations like natural disasters or when equipment breaks down. To help with this, people can use tricks like adding noise to the data or making fake data. This makes the training data more varied, which helps the AI models work better in new situations. Another issue is how tricky it is to fine-tune AI models for optical fiber sensor data. This needs a lot of computing power and know-how. To wrap up, AI-powered optical fiber sensing marks a big leap in how sensing tech has grown. By adding AI algorithms to optical fiber sensors, scientists and engineers can boost these systems' accuracy, flexibility, and usefulness. This opens doors to fresh possibilities across many industries. This study provides a guideline for the future studies for optimal planning of AI techniques selection and implementation, as shown in Fig. 1.

2 Sensor Data and AI Integration

Optical fiber sensor works on the fundamental principal of controlling the light intensity according to the variations in external physical factors like temperature, pressure, and strain. Traditionally, the data collected from these sensors has been analyzed using standard signal processing methods. However, the integration of AI is significantly transforming the way this data is processed and understood.

In optical fiber sensing, there are various kinds of data generated with different sensor types. These data can be classified based on physical or chemical quantity. They are used to measure the method of signal modulation. The common data types in optical fiber sensors are provided in the Table 1.

Intensity modulated data measures the change in the light intensity which are crucial in displacement detectors, and used to measure temperature, strain, pressure, etc. Wavelength-based sensors utilizes the shifts in the wavelength of the light, as used in Fiber Bragg Gratings. Phase-based sensors detect changes in the phase of the light waves traveling through the optical fiber sensors, and measures the conditions like pressure, temperature, or sound waves affecting the fiber. Polarization of the light for measuring stress and strain measurements uses polarization-based sensors.

For high-precision measurements, such as laser doppler velocimetry frequency-based sensors, are utilized which relies on the modulation of the frequency of light waves in the fiber due to external stimuli. These sensors are also utilized to evaluate the time difference in reaching optical fiber rays. These sensors are crucial in measuring the defects in the optical fiber cables which are important for their operation. Spectral-based sensors analyzes the change in the spectrum of light due to

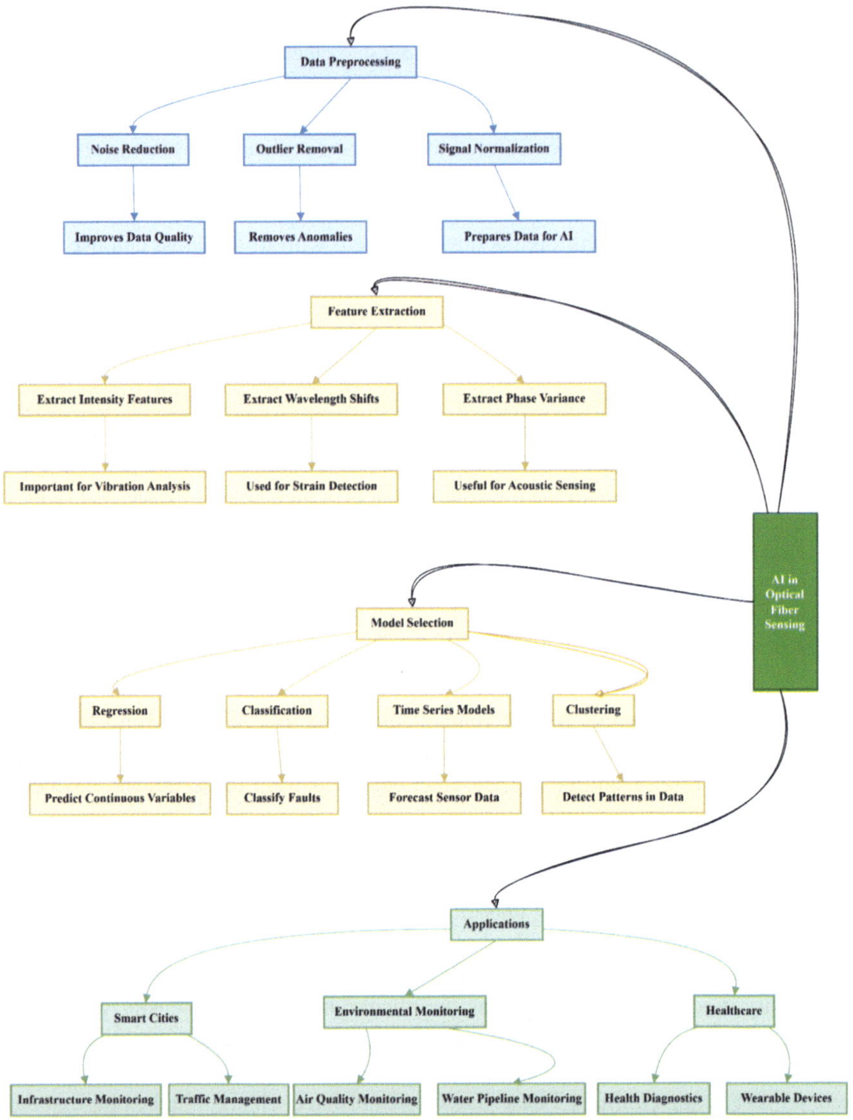

Fig. 1 Layout of the AI application in optical fiber sensing

absorption, scattering, or fluorescence. These sensors have application in chemical and gas industries to detect the leakage.

In optical sensor data analysis, pre-processing of the received data is very crucial for ML models. The performance of any machine learning or deep learning model heavily relies on the dataset. In pre-processing step, the data collected from the

Table 1 Optical fiber sensors data types

Data type	Measurement principle	Characteristics	Applications	Additional parameters for ML analysis
Intensity modulated	Measures changes in light intensity due to external effects	–Simple design	–Displacement sensors	–Intensity variation over time
		–Sensitive to bending, misalignment, and light loss	–Vibration monitoring	–Signal-to-noise ratio (SNR)
				–Noise characteristics
Wavelength modulated	Detects wavelength shifts (e.g., Fiber Bragg Gratings)	–Accurate and stable	–Strain and temperature sensors (FBG)	–Wavelength shift patterns
		–Not sensitive to light intensity fluctuations		–Slope and magnitude of shifts
				–Spectral resolution
Phase modulated	Measures phase shifts in light waves	–High sensitivity	–Acoustic monitoring	–Phase variance
		–Requires complex detection systems	–Seismic and pressure sensors	–Time delay changes
				–Phase noise
Polarization modulated	Detects changes in the polarization state of light	–Sensitive to stress and strain	–Stress and strain measurement	–Polarization state shifts
		–Requires polarization-maintaining fibers		–Stress-to-polarization ratio
				–Noise due to external perturbations
Frequency modulated	Measures frequency shifts or changes in light	–High precision and accuracy	–Laser Doppler velocimetry	–Frequency shift dynamics
		–Requires advanced modulation techniques	–High-precision sensing	–Doppler shift calculations
				–Frequency stability
Time-of-flight modulated	Measures time delay of light pulses traveling in the fiber	–Long-range distributed sensing	–Fault detection	–Time delay patterns
		–Reflects position of disturbances along fiber	–Distance and position sensing	–Signal attenuation
				–Path length variations
Spectral modulated	Detects changes in the spectrum, absorption, or scattering of light	–Sensitive to chemical or environmental changes	–Gas sensing	–Spectral intensity profile
		–High specificity	–Chemical and biological detection	–Absorption peaks
				–Signal dispersion characteristics

Table 2 Sensor data preprocessing steps

Data type	Original features	Preprocessing step	Output features	Comments
Intensity modulated	Intensity (I), time (t)	Normalization, smoothing	Normalized intensity	Address noise and drift
Wavelength modulated	Wavelength (λ), strain	Scaling, outlier removal	Normalized wavelength	Remove noisy readings
Phase modulated	Phase (φ), time (t)	Phase unwrapping, filtering	Filtered phase	Handle phase ambiguity
Polarization modulated	Polarization (P), stress	PCA, filtering	Reduced dimension	Reduce dimensionality
Frequency modulated	Frequency (f), time (t)	FFT, noise reduction	Frequency spectrum	Extract spectral features
Time-of-flight modulated	Time delay (τ), distance	Signal reconstruction	Time corrected signal	Correct signal losses
Spectral modulated	Spectrum (S), wavelength	Denoising, smoothing	Smoothed spectrum	Improve feature clarity

optical fiber sensors are checked for the noise, anomalies, and missing values. These variations in the data can occur due to different envoirmental or operation factors.

2.1 *Sensor Data Preprocessing*

Table 2 depicts different data types and possible pre-processing steps that could be adopted on these sensor datasets. In the pre-processing step the major operations are normalization, filtering, noise reduction, or phase unwrapping. The normalization of the sensor data converts every parameter on the same scale. This linearization of scale provides every feature variable equal opportunity in the training of the ML model. Filtering and noise reduction limit the anomalies in the dataset by removing very high or low sensor values which are normally not possible.

2.2 *Feature Extraction for AI Models*

The features are important for analyzing the data by extracting the meaning features from the sensor data. After cleaning the data with pre-processing step, the important features are extracted from the feature variables which basically transform raw data into actionable features.

The extracted features can be peak-to-peak amplitude, wavelength shift, or phase variance, which are crucial for model training, as shown in Table 3. These features are extracted to interpret the modeling better and for improving the prediction accuracy.

Table 3 Feature extraction methods on different sensor datasets

Data type	Raw features	Extracted features	Comments
Intensity modulated	Intensity (I), time (t)	Peak-to-peak amplitude, RMS value, Zero-crossing rate	Important for vibration analysis
Wavelength modulated	Wavelength (λ), strain	Wavelength shift, slope, strain coefficient	Key for temperature/strain sensing
Phase modulated	Phase (φ), time (t)	Phase variance, group delay	Useful for acoustic sensing
Polarization modulated	Polarization (P)	Degree of polarization, polarization rotation	Strain and stress indicator
Frequency modulated	Frequency (f), time (t)	Frequency shift, power spectral density	Extract Doppler shifts
Time-of-flight modulated	Time delay (τ), distance	Time of flight, Signal attenuation	Distance and fault location
Spectral modulated	Spectrum (S), wavelength	Spectral peaks, absorption bandwidth	Useful in chemical detection

These methods also reduce the computational burden by reducing the dimensions of the dataset.

2.3 *Data Augmentation Methods*

Data is an important tool in analyzing any task and predicting. Data generation with real-time systems are costly and time-consuming affair. Table 4 shows different data augmentation techniques to increase the dataset using AI methods. Data types in the Table 4 show the different sensor data types such as intensity-modulated, wavelength-modulated, phase modulated, and polarization modulated sensor datasets.

Table 4 Data augmentation methods for different optical fiber sensing data types

Data Type	Augmentation technique	Reason for augmentation	Comments
Intensity modulated	Add noise to signal, random jitter	Increase robustness to noise	Simulate sensor noise conditions
Wavelength modulated	Time shifting, frequency shifting	Generalize model to unseen conditions	Apply to simulate different environments
Phase modulated	Synthetic phase shift creation	Enhance small dataset size	Create variations of real data
Polarization modulated	Varying polarization angle	Test stress/strain effects in new scenarios	Useful for synthetic stress data

In intensity modulated data, adding noise to intensity data could simulate environmental interference, helping the model learn to filter noise in real applications. During the training phase of the model small noise is added, and model tries to identify and reduce the noise and predict the original data. In other sensor environment the time shifting, synthetic phase shift creation, and varying polarization angle method is utilized to train the model and correct the subsequent prediction.

3 Model Development and Optimization

The development in the field of AI has led to the availability and generation of numerous models. However, this large number of model availability makes the appropriate model selection a tough task. Moreover, any single model cannot perform better than other models in all kinds of datasets. This section of the chapter delves deep into the appropriate model selection with other parameter optimization for better performance of optical fiber sensing system.

3.1 Selection of Machine Learning Models

The analysis with AI model can be classified into three main categories such as regression, classification, and clustering. Table 5 shows the class of machine learning models that can be utilized with type of optical fiber senor datasets. Regression lies in supervised learning algorithm and used to predict a continuous output variable based on input features. Means in regression target variable is provided to guide the model for optimal function generation. Classification is also a type of supervised learning, but it is used for predicting discrete categories or classes. In classification the input is classified into predefined classes which are made during the training of input data. It is commonly used in fault detection in optical sensor data. Clustering lies in the unsupervised class of machine learning. During training there is not any target given to the machine learning model, it basically divides the dataset into different clusters. It is normally employed to check the pattern among different sensor datasets.

Table 5 Machine learning models for different optical fiber sensor data types

Model Type	Algorithm	Hyperparameters	Comments
Regression	Linear regression, polynomial regression	Regularization (L2), learning rate	Used for predicting intensity trends
Classification	Support vector machine (SVM)	Kernel type (RBF), C (Penalty), gamma	Applied to classify faults
Clustering	K-means clustering	Number of clusters (k)	Grouping sensor behavior patterns

Table 6 Performance metrices for machine learning models

Model type	Algorithm	Metrics
Regression	Linear regression	RMSE, MAE
Classification	Random forest classifier	Accuracy, F1 score, precision
Clustering	K-means	Accuracy

3.2 *Evaluation Metrics*

Machine learning model's performance are analyzed based on performance metrices. Table 5 provides the overview of metrices utilized in different classes of machine learning model's performance evaluation. Root Mean Square Error (RMSE) and Mean Absolute Error (MAE) are normally utilized to measure the performance of regression problems, whereas accuracy, F1 score, and precision are used to measure the performance of classification problems. In clustering, the grouping performance are normally measured with the help of model's accuracy in predicting the exact class of new data. So systematical organization of evaluation metrics in Table 6 helps to determine which model is best suited for the task and ensures that models are judged based on appropriate criteria for the dataset and problem.

4 Conclusion

AI is gaining momentum in different parts of optical fiber sensing application, unlocking new possibilities in precision, adaptability, and real-time data processing. This study has explored all the possible combination of options of different parts optical fiber sensing technology application. The optical fiber sensors have different kinds of data types in different fields of applications that show that data could vary from simple long-term dataset or high-resolution large datasets. These varying datasets will change the pre-processing steps adoption for cleaning the dataset and output for the training phase. The data generation for proper operation prediction and maintenance is a crucial step for an optimal operation. To address this limitation this, work also provides different options of data generation methods in data augmentation section. Moreover, the different machine learning models' selection according to the problem class is also discussed in detail, with suitable evaluation metrics for each class of the problem. This study could provide a foundation for utilizing the AI techniques in the optical fiber sensing techniques.

Acknowledgements We are deeply grateful to Dr. Amit Rai for his invaluable guidance and support throughout the preparation of this book chapter. His expertise and mentorship were instrumental in shaping our research and ensuring the quality of our work.

References

1. B. Lee, Review of the present status of optical fiber sensors. Opt. Fiber Technol. **9**, 57 (2003)
2. P. Lu, N. Lalam, M. Badar, B. Liu, B.T. Chorpening, M.P. Buric, P.R. Ohodnicki, Distributed optical fiber sensing: Review and perspective. Appl. Phys. Rev. **6**, (2019)
3. H.-E. Joe, H. Yun, S.-H. Jo, M.B.G. Jun, B.-K. Min, A review on optical fiber sensors for environmental monitoring. Int. J. Precis. Eng. Manuf. Green Technol. **5**, 173 (2018)
4. Y. Sun, J. Liu, Z. Xue, Q. Li, C. Fan, X. Zhang, A critical review of distributed fiber optic sensing for real-time monitoring geologic CO_2 sequestration. J. Nat. Gas. Sci. Eng. **88**, 103751 (2021)
5. M. Lyu, Z. Lin, G. Li, G. Situ, Fast modal decomposition for optical fibers using digital holography. Sci. Rep. **7**, 6556 (2017)
6. F. Liu, Q. Xu, Y. Liu, Strain data correction of distributed optical fiber sensors using strain transfer model with variable shear lag parameters. Autom. Constr. **140**, 104311 (2022)
7. S. Wang, Y. Yang, L. Zhang, L. Mohanty, R.-B. Jin, S. Wu, P. Lu, High-precision fiber optic liquid level sensor based on fast Fourier amplitude demodulation in a specific range of spectrum. Measurement **187**, 110326 (2022)
8. T. Jannson, Real-time Fourier transformation in dispersive optical fibers. Opt. Lett. **8**, 232 (1983)
9. C.M. De Lima Costa, M.E.B. Gonçalves, J.G.S. Ramalho, L.C. Da Silva, P.D.P. Dias, J.S. Almeida, D.L. Silva, V.L. Da Silva, Optical fiber vibration sensor for automated inspection of industrial assets, in *2023 Latin American Robotics Symposium (LARS), 2023 Brazilian Symposium on Robotics (SBR), and 2023 Workshop on Robotics in Education (WRE)* (IEEE, 2023), pp. 445–448
10. Y. Zhao, Y. Liu, B. Han, M. Wang, Q. Wang, Y. Zhang, Fiber optic volatile organic compound gas sensors: A review. Coord. Chem. Rev. **493**, 215297 (2023)
11. J. Nedoma, M. Fajkus, R. Martinek, J. Vanus, S. Kepak, R. Kahankova, R. Jaros, D. Cvejn, M. Prauzek, Analysis of the use of fiber-optic sensors in the road traffic. IFAC-PapersOnLine **51**, 420 (2018)
12. Y. Zhao, Z. Lin, S. Dong, M. Chen, Review of wearable optical fiber sensors: Drawing a blueprint for human health monitoring. Opt. Laser Technol. **161**, 109227 (2023)
13. X. Zhang, C. Wang, T. Zheng, H. Wu, Q. Wu, Y. Wang, Wearable optical fiber sensors in medical monitoring applications: A review. Sensors **23**, 6671 (2023)

The manufacturer's authorised representative in the EU is Springer Nature Customer Service Centre GmbH, Europaplatz 3, 69115 Heidelberg, Germany. If you have any concerns regarding our products, please contact ProductSafety@springernature.com

Printed and bound by CPI Group (UK) Ltd, Croydon, CR0 4YY
01/10/2026
02225922-0003